Math Contests
for Grades 4, 5, and 6
Volume 6

**School Years
2006-2007 through 2010-2011**

Written by

Steven R. Conrad • Daniel Flegler • Adam Raichel

Published by MATH LEAGUE PRESS
Printed in the United States of America

Cover art by Bob DeRosa

Phil Frank Cartoons Copyright © 1993 by CMS

First Printing, 2011

Copyright © 2011
by Mathematics Leagues Inc.
All Rights Reserved

No part of this publication may be reproduced or transmitted in any form or by any means, electronic or mechanical, including photocopy, recording, or any information storage or retrieval system, or any other means, without written permission from the publisher. Requests for permission or further information should be addressed to:

Math League Press
P.O. Box 17
Tenafly, NJ 07670-0017

ISBN 978-0-940805-18-7

Preface

Math Contests—Grades 4, 5, and 6, Volume 6 is the sixth volume in our series of problem books for grades 4, 5, and 6. The first five volumes contain the contests given in the school years 1979-1980 through 2005-2006. This volume contains contests given from 2006-2007 through 2010-2011. (You can use the order form on page 154 to order any of our 18 books.)

This book is divided into three sections for ease of use by students and teachers. You'll find the contests in the first section. Each contest consists of 30, 35, or 40 multiple-choice questions that you can do in 30 minutes. On each 3-page contest, the questions on the 1st page are generally straightforward, those on the 2nd page are moderate in difficulty, and those on the 3rd page are more difficult. In the second section of the book, you'll find detailed solutions to all the contest questions. In the third and final section of the book are the letter answers to each contest. In this section, you'll also find rating scales you can use to rate your performance.

Many people prefer to consult the answer section rather than the solution section when first reviewing a contest. We believe that reworking a problem when you know the answer (but *not* the solution) often leads to increased understanding of problem-solving techniques.

Each year we sponsor an Annual 4th Grade Mathematics Contest, an Annual 5th Grade Mathematics Contest, and an Annual 6th Grade Mathematics Contest. A student may participate in the contest on grade level or for any higher grade level. For example, students in grades 4 and 5 (or below) may participate in the 6th Grade Contest. Starting with the 1991-92 school year, students have been permitted to use calculators on any of our contests.

Steven R. Conrad, Daniel Flegler, & Adam Raichel, contest authors

Acknowledgments

For her continued patience and understanding, special thanks to Marina Conrad, whose only mathematical skill, an important one, is the ability to count the ways.

For demonstrating the meaning of selflessness on a daily basis, special thanks to Grace Flegler.

To Jeannine Kolbush, who did an awesome proofreading job, thanks!

Table Of Contents

Preface i

Acknowledgements ii

Grade	School Year	Page for Contest	Page for Solutions	Page for Answers
4	2006-07	5	73	138
4	2007-08	9	77	139
4	2008-09	13	81	140
4	2009-10	17	85	141
4	2010-11	21	89	142
5	2006-07	27	95	143
5	2007-08	31	99	144
5	2008-09	35	103	145
5	2009-10	39	107	146
5	2010-11	43	111	147
6	2006-07	49	117	148
6	2007-08	53	121	149
6	2008-09	57	125	150
6	2009-10	61	129	151
6	2010-11	65	133	152

Order Form For Contest Books (Grades 4-12) 154

The Contests

• •

2006-2007 through 2010-2011

4th Grade Contests
∙∙∙∙∙∙∙∙∙∙∙∙∙∙∙∙∙∙∙∙∙∙∙∙∙
2006-2007 through 2010-2011

FOURTH GRADE MATHEMATICS CONTEST

Math League Press, P.O. Box 17, Tenafly, New Jersey 07670-0017

2006-2007 Annual 4th Grade Contest

Spring, 2007

4

Instructions

Time You will have only *30 minutes* working time for this contest. You might be *unable* to finish all 30 questions in the time allowed.

Scores Please remember that *this is a contest, not a test*—and there is no "passing" or "failing" score. Few students score as high as 24 points (80% correct). Students with half that, 12 points, *deserve commendation!*

Format and Point Value This is a multiple-choice contest. Each answer is an A, B, C, or D. Write each answer in the *Answer Column* to the right of each question. A correct answer is worth 1 point. Unanswered questions get no credit. You **may** use a calculator.

Copyright © 2007 by Mathematics Leagues Inc.

2006-2007 4TH GRADE CONTEST

1. Of the following, which has the largest value?
   A) 3 + 3   B) 3 − 3   C) 3 × 3   D) 3 ÷ 3

2. Twelve days after a Monday is a
   A) Wednesday B) Thursday C) Friday D) Saturday

3. 55 + 55 + 55 = 44 + 44 + ?
   A) 33   B) 44   C) 66   D) 77

4. How many 5¢ stamps can I buy with 5 dimes?
   A) 10   B) 20   C) 25   D) 50

5. Each of The Chef's omelets uses 3 eggs. How many omelets can The Chef make with 1 dozen brown eggs, 1 dozen white eggs, and 1 dozen speckled eggs?
   A) 1 dozen   B) 2 dozen
   C) 3 dozen   D) 4 dozen

6. 8 × 9 × 10 × 11 = 80 × ?
   A) 81   B) 88   C) 90   D) 99

7. In which of the following divisions is 2 the remainder?
   A) 463 ÷ 1   B) 874 ÷ 2   C) 355 ÷ 3   D) 506 ÷ 4

8. The word mathematics has twice as many vowels as the word
   A) arithmetic   B) history   C) reading   D) computers

9. Eight pairs of bears have ? pairs of paws.
   A) 16   B) 32   C) 48   D) 64

10. 48 ÷ 4 = 4 × ?
    A) 3   B) 8   C) 12   D) 16

11. If 19 policemen are on patrol in 9 cars, and each car is used by at least 2 policemen, then exactly 1 patrol car is used by ? policemen.
    A) 2   B) 3   C) 4   D) 5

Go on to the next page ▐▶ **4**

2006-2007 4TH GRADE CONTEST

12. Which of the following products is an odd number?
    A) 23×24   B) 24×35   C) 42×53   D) 53×45

13. Together, Pat and I ate 24 sandwiches. If I ate 6 more sandwiches than Pat, then I ate ? sandwiches?
    A) 9   B) 12   C) 15   D) 18

14. Subtract 20 from 4 times ? to get 100.
    A) 120   B) 80   C) 30   D) 25

15. Suzie sold 7 fewer than 7×70 seashells at the seashore. How many seashells did Suzie sell at the seashore?
    A) 7×69   B) 7×63   C) 6×70   D) 6×63

16. Mary's brother has 3 sisters, so Mary has ? sisters.
    A) 2   B) 3   C) 4   D) 5

17. In total, ? triangles have as many sides as 12 squares.
    A) 4   B) 16   C) 36   D) 48

18. My sunflower doubles its height every day. It was 10 cm tall at noon yesterday. It will be ? cm tall at noon the day after tomorrow.
    A) 30   B) 40   C) 60   D) 80

19. Which leaves the greatest remainder when divided by 5?
    A) 543   B) 654   C) 876   D) 987

20. 4×8×16 = ? ×16×16
    A) 2   B) 4   C) 8   D) 16

21. Al milked Betsy from 3:18 P.M. to 4:56 P.M. today. At ? , Al had finished half his milking time.
    A) 3:58 P.M.   B) 4:04 P.M.
    C) 4:07 P.M.   D) 4:19 P.M.

22. A number divisible by both 6 and 20 must also be divisible by
    A) 12   B) 14   C) 26   D) 120

Go on to the next page ⏵ 4

	2006-2007 4TH GRADE CONTEST	Answer Column

23. The product of 6 of the numbers 1, 2, 3, 4, 5, 6, 7, 8, 9 could be

    A) 680   B) 2835
    C) 60 480   D) 83 349

    23.

24. My 3 pets birds are brothers. The sum of their ages is 25. In 2 years, that sum will be

    A) 27   B) 29   C) 30   D) 31

    24.

25. A *prime* is a whole number greater than 1 that's divisible by only 1 and itself. The greatest prime number less than 50 is

    A) 49   B) 47   C) 41   D) 37

    25.

26. The length of a certain rectangle is twice its width. If the perimeter of the rectangle is 24, its width is

    A) 2   B) 3   C) 4   D) 6

    26.

27. If my florist needs 15 flowers to make 3 special bouquets, then how many flowers would my florist need to make 18 special bouquets?

    A) 90   B) 54   C) 45   D) 36

    27.

28. If I have as many pennies as dimes, then the total value of all these coins could be

    A) $1.01   B) $1.10
    C) $1.11   D) $11.01

    28.

29. Triangles are *different* unless they share 3 vertices. Exactly ? different triangles of all shapes and sizes can be traced along the lines already in the figure.

    A) 4   B) 6   C) 8   D) 12

    29.

30. (100 + 98 + 96 + ... +2) − (99 + 97 + 95 + ... + 1) =

    A) 25   B) 50   C) 99   D) 100

    30.

The end of the contest 4

Visit our Web site at http://www.mathleague.com

Solutions on Page 73 • Answers on Page 138

FOURTH GRADE MATHEMATICS CONTEST

Math League Press, P.O. Box 17, Tenafly, New Jersey 07670-0017

2007-2008 Annual 4th Grade Contest

Spring, 2008

Instructions

4

- **Time** You will have only *30 minutes* working time for this contest. You might be *unable* to finish all 30 questions in the time allowed.

- **Scores** Please remember that *this is a contest, not a test*—and there is no "passing" or "failing" score. Few students score as high as 24 points (80% correct). Students with half that, 12 points, *deserve commendation!*

- **Format and Point Value** This is a multiple-choice contest. Each answer is an A, B, C, or D. Write each answer in the *Answer Column* to the right of each question. A correct answer is worth 1 point. Unanswered questions get no credit. You **may** use a calculator.

Copyright © 2008 by Mathematics Leagues Inc.

2007-2008 4TH GRADE CONTEST

1. $2 \times 0 \times 0 \times 8 =$
   A) 2008 B) 16 C) 10 D) 0

2. 4+4+4 + 4+4+4 has the same value as
   A) 1×18 B) 2×12
   C) 2×16 D) 4×8

3. 444 + 555 = 666 + ?
   A) 222 B) 333 C) 444 D) 777

4. If tomorrow is Tuesday, then 2 days before yesterday was
   A) Friday B) Saturday C) Sunday D) Monday

5. The product of 3, 3, 3, and 3 is
   A) 12 B) 27 C) 81 D) 3333

6. $4 \times 1 \times 3 \times 1 \times 2 \times 1 =$
   A) 9 B) 12 C) 24 D) 39

7. Forty panda bears have ? pairs of front paws.
   A) 20 B) 40 C) 80 D) 160

8. 91−19 = 81− ?
   A) 9 B) 18 C) 19 D) 29

9. Of the 24 bears who live in Yogi National Park, 5 were seated at a picnic table and ? were not.
   A) 5 B) 9
   C) 15 D) 19

10. $(99 \div 3) \times 3 =$
    A) 11 B) 33 C) 99 D) 297

11. $(10-10) \times (10+10) =$
    A) 0 B) 10 C) 20 D) 100

12. Of the following, which is a whole number?
    A) 242 ÷ 4 B) 422 ÷ 4 C) 442 ÷ 4 D) 452 ÷ 4

13. There are an odd number of days in ? weeks.
    A) 66 B) 44 C) 33 D) 22

Go on to the next page ⮕ **4**

2007-2008 4TH GRADE CONTEST

14. The ones' digit of $99 \times 91 \times 19 \times 11 \times 9$ is
    A) 9 B) 7 C) 3 D) 1

15. Lionel's train delivers supplies to the top of a mountain. The entire train is only 4 times as long as its engine. If the entire train is 120 m long, then the length of the engine is ? m.
    A) 30 B) 80 C) 124 D) 480

16. I counted 600 legs on the dogs in a dog show. How many dogs were in the show?
    A) 120 B) 150 C) 180 D) 300

17. Of the following, which is *not* a prime number?
    A) 11 B) 31 C) 41 D) 51

18. To solve the equation, replace each word with the number less than 10 that rhymes with it:
    (zoo × kicks) + (door × gate) = ?
    A) 20 B) 32 C) 44 D) 64

19. *Palindrome* years (like 1991 and 2002) read the same forwards and backwards. The years 1991 and 2002 are 11 years apart. How many years apart are 2002 and the first *palindrome* year after 2002?
    A) 11 B) 110 C) 111 D) 220

20. I like to make snowmen. Each time I roll a snowball, its radius doubles. If its radius is now 10 cm long, then 4 rolls from now its radius will be ? cm long.
    A) 40 B) 50 C) 80 D) 160

21. If 3 apples weigh as much as 4 pears, and 2 pears weigh as much as 5 plums, then 9 apples weigh as much as ? plums.
    A) 10 B) 19 C) 27 D) 30

22. What time is it 62 minutes before 15 hours after midnight?
    A) 1:58 P.M. B) 2:58 P.M. C) 3:58 P.M. D) 5:58 P.M.

Go on to the next page ⇒ **4**

2007-2008 4TH GRADE CONTEST

Answer Column

23. Each time a chartered bus with 18 people stops, 3 people get off and then 2 get on. This continues until the bus is empty. That takes ? stops.
    A) 18   B) 16
    C) 9   D) 6

23.

24. If the length of each side of a triangle is an odd number, then the perimeter of the triangle might be
    A) 19   B) 20   C) 22   D) 24

24.

25. Starting at 2, counting to 200 by 2s, the 99th number counted is
    A) 100   B) 102   C) 192   D) 198

25.

26. Among 6 coins worth 19¢, there are ? more pennies than dimes.
    A) 1   B) 2   C) 3   D) 4

26.

27. How many 2-digit whole numbers have no odd factor except 1?
    A) 49   B) 45   C) 3   D) 0

27.

28. A room has 24 couch potatoes and 18 couches. At least 1 couch potato sits on every couch. At most ? couches have exactly 2 couch potatoes sitting on them.
    A) 9 B) 6 C) 4 D) 3

28.

29. If the length of each side of a pentagon is the same even number, then the pentagon's perimeter could be
    A) 210   B) 444   C) 666   D) 864

29.

30. If $1 + 2 + \ldots + 50 = 1275$, then $5 + 10 + \ldots + 250 =$
    A) 1525   B) 1775   C) 3825   D) 6375

30.

The end of the contest 4

Visit our Web site at http://www.mathleague.com

Solutions on Page 77 • Answers on Page 139

FOURTH GRADE MATHEMATICS CONTEST

Math League Press, P.O. Box 17, Tenafly, New Jersey 07670-0017

2008-2009 Annual 4th Grade Contest

Spring, 2009

Instructions

4

- **Time** You will have only *30 minutes* working time for this contest. You might be *unable* to finish all 30 questions in the time allowed.

- **Scores** Please remember that *this is a contest, and not a test*—there is no "passing" or "failing" score. Few students score as high as 24 points (80% correct). Students with half that, 12 points, *deserve commendation!*

- **Format and Point Value** This is a multiple-choice contest. Each answer is an A, B, C, or D. Write each answer in the *Answer Column* to the right of each question. A correct answer is worth 1 point. Unanswered questions receive no credit. You **may** use a calculator.

Copyright © 2009 by Mathematics Leagues Inc.

2008-2009 4TH GRADE CONTEST

1. What is 6 more than the sum of 5 and 4?

   A) 11   B) 13   C) 15   D) 96

2. If Alice has a prime number of sisters, Alice could have _?_ sisters.

   A) 6   B) 7   C) 8   D) 9

3. $8 \times 9 \times 0 =$

   A) 0   B) 17   C) 72   D) 890

4. What is 216 rounded to the nearest 10?

   A) 200   B) 210   C) 220   D) 300

5. $33 + 33 + 33 = 3 + 3 + 3 +$ _?_

   A) 3   B) 30   C) 90   D) 99

6. If Blake left for school 30 minutes after 7:50 A.M., he left at

   A) 8:00 A.M.   B) 8:20 A.M.   C) 8:30 A.M.   D) 8:50 A.M.

7. $15 - 10 + 25 - 20 + 35 - 30 + 45 - 40 =$

   A) 5   B) 10   C) 15   D) 20

8. 2 dollars + 2 quarters + 2 dimes + 2 nickels = _?_ pennies

   A) 280   B) 250   C) 222   D) 215

9. $(10 + 30 + 50 + 70) - (20 + 40 + 60) =$

   A) 40   B) 50   C) 60   D) 70

10. It takes Carlos 15 minutes each time he washes his dog. If he spent 3 hours total washing his dog in March, how many times did he wash his dog that month?

    A) 2   B) 4   C) 8   D) 12

11. $18 \times 16 \times 14 = 9 \times 8 \times 7 \times$ _?_

    A) 2   B) 4   C) 6   D) 8

Go on to the next page))))▶ **4**

2008-2009 4TH GRADE CONTEST

12. Of the following, which is not equal to $2 \times 3 \times 4 \times 6$?

    A) 6×24 B) 8×9 C) 12×12 D) 8×18

13. The number of golf balls Eric carries when he plays golf is always divisible by both 12 and 32. At least how many golf balls does Eric carry when he plays golf?

    A) 4 B) 44 C) 96 D) 384

14. $2 \times 3 \times 4 \times 5 \times 6 \times 7 = 6 \times 7 \times 8 \times 9 \times 10 \div \underline{\ ?\ }$

    A) 4 B) 5 C) 6 D) 7

15. What time is it seven hours and six minutes after 7:06 A.M.?

    A) 4:12 A.M. B) 12:00 P.M. C) 2:12 P.M. D) 12:04 A.M.

16. There are 6 people ahead of Faye and 6 behind her as she stands in line. All together how many people are in line?

    A) 10 B) 11 C) 12 D) 13

17. If you triple a number and divide the result by 5, the quotient is 6. What number are you tripling?

    A) 8 B) 10 C) 11 D) 15

18. What is the sum of the number of sides in a square, the number of sides in a rectangle, and the number of sides in a parallelogram?

    A) 11 B) 12 C) 13 D) 14

19. 16 less than 38 = 11 more than $\underline{\ ?\ }$

    A) 11 B) 16 C) 38 D) 65

20. Doubling an even number and then adding 8 always results in a multiple of

    A) 4 B) 5 C) 6 D) 7

21. Diana puts socks on the 4 paws of each cat living on her street. If she uses 2 dozen pairs of socks, on exactly how many cats does she put socks?

    A) 4 B) 6 C) 12 D) 24

Go on to the next page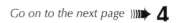

2008-2009 4TH GRADE CONTEST

22. 135 798 642 is *not* divisible by

    A) 3   B) 4   C) 6   D) 9

23. The product of the number of pets I have and the number of pets my friend has is 24. The total number of pets we have *cannot* be

    A) 25   B) 11   C) 10   D) 8

24. Each player on a football team wears a jersey with a two-digit number between 10 and 99 on it, and each jersey number is a different multiple of 4. At most how many players are on the team?

    A) 9   B) 18   C) 22   D) 24

25. What is the perimeter of a rectangle if the sum of its length and width is 13?

    A) 13   B) 26   C) 39   D) 52

26. Ivan has 4 times as many stamps as Jackie, who has 3 times as many as Kenji. If Ivan has 36 stamps, how many does Kenji have?

    A) 3   B) 4   C) 9   D) 12

27. Lee's house is 4 km from Mary's and 7 km from Nat's. What is the greatest possible distance between Mary's house and Nat's house?

    A) 3 km   B) 4 km   C) 7 km   D) 11 km

28. When 216 is divided by __?__ , the remainder is 4.

    A) 53   B) 34   C) 20   D) 4

29. Gary is three times as old as Harry will be 4 years from now. If Harry is 5 years old now, how old will Gary be in 4 years?

    A) 10   B) 23   C) 27   D) 31

30. A camp lets each camper choose 2 of 6 available activities. If there are 180 campers, what is the average number of campers in each activity?

    A) 30   B) 60   C) 120   D) 180

The end of the contest 4

Visit our Web site at http://www.mathleague.com
Solutions on Page 81 • Answers on Page 140

FOURTH GRADE MATHEMATICS CONTEST

Math League Press, P.O. Box 17, Tenafly, New Jersey 07670-0017

2009-2010 Annual 4th Grade Contest

Spring, 2010

Instructions

4

- **Time** Do *not* open this booklet until you are told by your teacher to begin. You will have only *30 minutes* working time for this contest. You might be *unable* to finish all 30 questions in the time allowed.

- **Scores** Please remember that *this is a contest, and not a test* — there is no "passing" or "failing" score. Few students score as high as 24 points (80% correct). Students with half that, 12 points, *should be commended!*

- **Format and Point Value** This is a multiple-choice contest. Each answer is an A, B, C, or D. Write each answer in the *Answer Column* to the right of each question. A correct answer is worth 1 point. Unanswered questions receive no credit. You **may** use a calculator.

Copyright © 2010 by Mathematics Leagues Inc.

2009-2010 4TH GRADE CONTEST

1. 2 + 0 + 1 + 0 = 2 + 1 + _?_

   A) 3   B) 2   C) 1   D) 0

2. I was 5 years old 4 years ago. How old will I be 3 years from now?

   A) 8   B) 9   C) 12   D) 13

3. There are 3 dozen dinosaur eggs in a nest. How many eggs is that?

   A) 15   B) 36   C) 42   D) 72

4. The product of 7 and 123 is

   A) 116   B) 130   C) 741   D) 861

5. Of the following, which is *not* a whole number?

   A) 3 + 2   B) 3 − 2   C) 3 × 2   D) 3 ÷ 2

6. Which number is two less than one more than ten thousand?

   A) 10 002   B) 10 001   C) 9999   D) 9998

7. Of the following, which product is an odd number?

   A) 212 × 12   B) 12 × 121   C) 2 × 1212   D) 1 × 2121

8. If today is Tuesday, 20 days from now will be a

   A) Monday   B) Tuesday   C) Wednesday   D) Saturday

9. Arnold the Giant Dog is holding a square board that has a perimeter of 12 m. How long is one side of the board?

   A) 2 m   B) 3 m   C) 4 m   D) 5 m

10. 10 dimes = _?_ nickels

    A) 20   B) 12   C) 5   D) 2

11. How many whole numbers are greater than 10 and less than 30?

    A) 18   B) 19   C) 20   D) 21

Go on to the next page))))➡ 4

2009-2010 4TH GRADE CONTEST

	Answers
12. 40 + 80 + 120 = 4 × ? A) 60 B) 50 C) 40 D) 30	12.
13. I have 1 dollar, 2 quarters, 3 dimes, and 4 nickels. This money is worth the same amount as ? pennies. A) 150 B) 175 C) 184 D) 200	13.
14. 37 hours after 6 o'clock is ? o'clock. A) 1 B) 4 C) 7 D) 12	14.
15. An elf's height increases by 10 cm each year. A gnome's height increases by half that amount each year. If the elf and the gnome were the same height exactly 2 years ago, how much taller than the gnome will the elf be exactly 2 years from today? A) 10 cm B) 15 cm C) 20 cm D) 25 cm	15.
16. What is the ones digit of (3 × 6) + (20 × 6) + (100 × 6)? A) 2 B) 4 C) 6 D) 8	16.
17. Charlie's 9th birthday is in 2010. In what year will he turn 21? A) 2019 B) 2020 C) 2021 D) 2022	17.
18. Multiplying a whole number by 4 and adding 6 will always result in a number that is A) even B) odd C) prime D) less than 100	18.
19. (36 + 36 + 36) ÷ 6 = 6 × ? A) 3 B) 4 C) 6 D) 9	19.
20. Darren dropped 6 of his flowers, but held on to 5 times as many flowers as he dropped. How many flowers did he have before he dropped any? A) 30 B) 36 C) 42 D) 56	20.
21. There are 1000 people in a stadium. If 87 are wearing hats, how many are *not* wearing hats? A) 813 B) 913 C) 987 D) 1087	21.

Go on to the next page))) 4

2009-2010 4TH GRADE CONTEST

22. 21 + 32 + 43 + 54 + 65 = 11 + 22 + 33 + 44 + 55 + _?_

    A) 20   B) 30   C) 40   D) 50

23. What is the sum of the digits of the number one million one hundred thousand one hundred one?

    A) 4   B) 6   C) 7   D) 8

24. Of the following numbers, which leaves the least remainder when divided by 9?

    A) 185   B) 280   C) 375   D) 470

25. The sum of the number of sides of a hexagon, a pentagon, and a rectangle is the same as the sum of the number of sides of _?_ triangles.

    A) 2   B) 3   C) 4   D) 5

26. At most how many non-overlapping squares with a perimeter of 8 can fit into a rectangle with width 5 and length 12?

    A) 5   B) 12   C) 15   D) 18

27. Each of five brothers was born on the same date but in different years, and today is their birthday. If each is an even number of years old today but each is less than 40 years old, what is the greatest possible sum of their ages?

    A) 38   B) 150   C) 170   D) 180

28. A farm has 4 henhouses loaded with chickens. Each henhouse has 6 rows of chickens, and there are 12 chickens per row. In all, how many chickens are there on that farm?

    A) 22   B) 24   C) 72   D) 288

29. Erin has fewer than 20 presents in her stack. If Erin divides 30 by the number of presents she has, the remainder *cannot* equal

    A) 16   B) 12   C) 8   D) 4

30. What is the difference between the sum of all whole numbers greater than 20 and less than 40, and the sum of all whole numbers greater than 40 and less than 60?

    A) 190   B) 200   C) 380   D) 400

The end of the contest 4

Visit our Web site at http://www.mathleague.com
Solutions on Page 85 • Answers on Page 141

FOURTH GRADE MATHEMATICS CONTEST

Math League Press, P.O. Box 17, Tenafly, New Jersey 07670-0017

2010-2011 Annual 4th Grade Contest

Spring, 2011

4

Instructions

- **Time** Do *not* open this booklet until you are told by your teacher to begin. You will have only *30 minutes* working time for this contest. You might be *unable* to finish all 30 questions in the time allowed.

- **Scores** Please remember that *this is a contest, and not a test*—there is no "passing" or "failing" score. Few students score as high as 24 points (80% correct). Students with half that, 12 points, *should be commended!*

- **Format and Point Value** This is a multiple-choice contest. Each answer is an A, B, C, or D. Write each answer in the *Answer Column* to the right of each question. A correct answer is worth 1 point. Unanswered questions receive no credit. You **may** use a calculator.

Copyright © 2011 by Mathematics Leagues Inc.

2010-2011 4TH GRADE CONTEST

1. The product of 2, 0, 1, and 1 is

   A) 0   B) 1   C) 2   D) 4

2. The sum of all even numbers between 1 and 9 is

   A) 8   B) 12   C) 20   D) 35

3. Didi the dancing dog danced for exactly one week. For how many hours did Didi dance?

   A) 24   B) 84   C) 128   D) 168

4. How many 25¢ gumballs can I buy for $5?

   A) 10   B) 20   C) 40   D) 80

5. I have 60 marbles. If I put as many of these marbles as I can into 7 equal piles, how many marbles are left over?

   A) 1   B) 4   C) 6   D) 11

6. Which number is 1000 tens minus 100 tens?

   A) 9000   B) 9900   C) 9990   D) 9999

7. I round 4567 to the nearest 10, then I subtract 4567. I end up with

   A) 3   B) 33   C) 333   D) 5433

8. If the perimeter of a rectangle is 20 and two of its sides each have a length of 4, what is the length of each of the other two sides?

   A) 16   B) 12   C) 8   D) 6

9. Otto did a handstand, and 1 out of every 5 pennies he had in his pockets fell out. If a total of 19 pennies fell out, Otto must have started with _?_ pennies in his pockets.

   A) 24   B) 76   C) 95   D) 114

10. I want to take a 75-minute nap starting at 5:15. The alarm used to end my nap should be set for

    A) 6:00   B) 6:15   C) 6:30   D) 6:45

11. The product of 4 odd numbers is always

    A) even   B) odd   C) prime   D) greater than 20

Go on to the next page)))➡ **4**

2010-2011 4TH GRADE CONTEST

12. The Bears' picnic has too many ants! If the number of ants is the least whole number greater than 0 that is divisible by both 12 and 42, there are ? ants.

    A) 54   B) 84   C) 124   D) 142

13. 15 + 15 + 15 + 15 +15 + 15 = 5 × _?_

    A) 18   B) 15   C) 6   D) 5

14. A zoo has 3 times as many female animals as male animals. If there are a total of 24 animals in the zoo, how many of these animals are female?

    A) 21   B) 18   C) 16   D) 6

15. Each side of my triangle is twice as long as the diameter of my circle. If the radius of my circle is 4, what is the perimeter of my triangle?

    A) 6   B) 24   C) 32   D) 48

16. $100 + $20 + $3 + 400¢ + 50¢ + 6¢ =

    A) $123.46   B) $124.56   C) $126.96   D) $127.56

17. How many whole numbers have more than 1 digit but fewer than 4?

    A) 986   B) 988   C) 990   D) 992

18. Maddie earned $8 per hour helping her friend move. She worked from Monday to Friday, 9 A.M. to 5 P.M. each day. How much did Maddie earn?

    A) $40   B) $64   C) $320   D) $448

19. There are 5 coins in my pocket. The total value of the coins could *not* be

    A) 14¢   B) 32¢   C) 43¢   D) 61¢

20. Of the following numbers, which is divisible by the greatest number of different whole numbers?

    A) 112   B) 90   C) 30   D) 8

21. $1 in quarters and $5 in nickels is _?_ coins.

    A) 8   B) 29   C) 104   D) 125

Go on to the next page ⟶ 4

2010-2011 4TH GRADE CONTEST

22. Chet made 80 pancakes. Some fell on plates, but the rest fell on his face. If the number of pancakes on plates is 16 greater than the number on his face, _?_ are on plates.

    A) 24   B) 32   C) 48   D) 56

23. Any number that is divisible by both 12 and 5 must also be divisible by

    A) 21   B) 15   C) 9   D) 8

24. When 123 is divided by 4, the sum of the quotient and remainder is

    A) 18   B) 24   C) 30   D) 33

25. The sum of three different even whole numbers *cannot* be

    A) 4   B) 12   C) 16   D) 20

26. The difference between the two largest divisors of 28 is

    A) 27   B) 14   C) 7   D) 6

27. There are two prime numbers between

    A) 10 and 15   B) 20 and 25   C) 26 and 30   D) 45 and 50

28. The first four stripes on a wall with 100 stripes are red, blue, white and purple, in that order. These four colors keep repeating in the same order. What color is the 55th stripe on the wall?

    A) red   B) blue   C) white   D) purple

29. In 20 years, Li will be 3 times as old as he is now. How old will he be in 10 years?

    A) 10   B) 20   C) 30   D) 50

30. Harry's beard was 4 cm long at the end of the day 4 days ago. If his beard doubles in length every 24 hours, how long will Harry's beard be at the end of the day tomorrow?

    A) 16 cm   B) 32 cm   C) 64 cm   D) 128 cm

The end of the contest 4

Visit our Web site at http://www.mathleague.com
Solutions on Page 89 • Answers on Page 142

5th Grade Contests
2006-2007 through 2010-2011

FIFTH GRADE MATHEMATICS CONTEST

Math League Press, P.O. Box 17, Tenafly, New Jersey 07670-0017

2006-07 Annual 5th Grade Contest

Spring, 2007

Instructions

5

- **Time** Do *not* open this booklet until you are told by your teacher to begin. You will have only *30 minutes* working time for this contest. You might be *unable* to finish all 30 questions in the time allowed.

- **Scores** Please remember that *this is a contest, not a test*—and there is no "passing" or "failing" score. Few students score as high as 24 points (80% correct). Students with half that, 12 points, *should be commended!*

- **Format and Point Value** This is a multiple-choice contest. Each answer is an A, B, C, or D. Write each answer in the *Answer Column* to the right of each question. A correct answer is worth 1 point. Unanswered questions get no credit. You **may** use a calculator.

Copyright © 2007 by Mathematics Leagues Inc.

2006-2007 5TH GRADE CONTEST

	Answer Column
1. 15 − 14 + 13 − 12 + 11 − 10 = A) 0 B) 1 C) 3 D) 6	1.
2. My class has 20 boys and twice as many girls. That's a total of A) 10 students B) 30 students C) 40 students D) 60 students	2.
3. Which of the following sums is divisible by 4? A) 1+1+1+1+1 B) 1+2+1+2+1 C) 2+1+2+1+2 D) 2+2+2+2+2	3.
4. One-third of 48 equals the product of 2 and A) 2 B) 4 C) 8 D) 16	4.
5. If it costs 75¢ to airmail a postcard from the United States to Australia, how much does it cost to mail 50 such postcards? A) $37.50 B) $50 C) $75 D) $3750	5.
6. 24 × 60 minutes = A) 1 day B) 60 hours C) 24 days D) 60 days	6.
7. A _?_ is worth a nickel less than a quarter more than a nickel. A) penny B) nickel C) dime D) quarter	7.
8. If Al's age is divided by 3, the remainder is 2. Al's age could be A) 13 B) 23 C) 33 D) 43	8.
9. At a campfire, 30 Scouts split into groups of 3. In each group, each Scout toasted 1 marshmallow for each of the other 2 Scouts. Altogether, how many marshmallows were toasted? A) 30 B) 60 C) 90 D) 120	9.
10. Of the following numbers, which is divisible by both 10 and 100? A) 1 010 010 B) 101 010 C) 100 010 D) 10 100	10.

Go on to the next page ⟶ **5**

2006-2007 5TH GRADE CONTEST

11. How many whole number factors of 27 are also factors of 72?
    A) 9 B) 4 C) 3 D) 2

12. Of the following, only ? is divisible by 7.
    A) 777 352 B) 770 143 C) 742 001 D) 728 714

13. Of my 12 cats, 3 times as many are males as females, so ? are males.
    A) 9 B) 8 C) 4 D) 3

14. At a rate of 72 photos in 2 hours, I can process 216 photos in
    A) 3 hours B) 6 hours
    C) 9 hours D) 12 hours

15. $27 \times 27 \times 27 = 9 \times 9 \times 9 \times$?
    A) 3 B) 9 C) 27 D) 81

16. Four friends and I split 60 coins equally. We each got ? coins.
    A) 20 B) 15 C) 12 D) 10

17. Class began at 1:23 PM, was half over at ?, and ended at 3:21 PM.
    A) 2:22 PM B) 2:24 PM C) 2:29 PM D) 2:31 PM

18. I need ? pages to print 6 three-page notes and 8 two-page notes.
    A) 6 B) 14 C) 34 D) 48

19. $(11 \times 12 \times 13 \times 14) \div (1 \times 2 \times 3 \times 4) =$
    A) $6 \times 11 \times 13$ B) $7 \times 11 \times 13$
    C) $11 \times 13 \times 14$ D) $10 \times 10 \times 10 \times 10$

    Things moving too fast for you?

20. If my pet turtle walks 60 cm every day, it walks ? in 20 days.
    A) 12 cm B) 30 cm
    C) 120 cm D) 1200 cm

21. ? pencils, each 4 cm long, can form a square of perimeter 48 cm.
    A) 3 B) 4 C) 8 D) 12

Go on to the next page ⏩ **5**

2006-2007 5TH GRADE CONTEST

22. A radius of each small circle is 4 cm long. How long is a radius of the large circle?
    A) 6 cm   B) 8 cm   C) 12 cm   D) 24 cm

23. If the sum of my age now and my age 5 years ago is 27, then my age 3 years ago was
    A) 13   B) 15   C) 19   D) 24

24. How many of the digits in 123 456 789 are factors of 123 456 789?
    A) 0   B) 3   C) 5   D) 7

25. The number ? is the product of 2 consecutive whole numbers.
    A) 597   B) 598   C) 599   D) 600

26. Ann hiccups 10 times every 3 minutes. Al sneezes 5 times every 2 minutes. Each hour, the number of times Al sneezes is ? less than the number of times Ann hiccups.
    A) 5   B) 10   C) 50   D) 100

27. If the sum of two whole numbers is 12, and half their product is 16, then their difference is
    A) 6   B) 4   C) 2   D) 0

28. What is the greatest odd number that is a divisor of $3 \times 6 \times 9$?
    A) 3   B) 9   C) 3×9 D) 9×9

29. Exactly how many different triangles actually appear in the star shown? (Any size is OK.)
    A) 10   B) 9   C) 8   D) 7

30. $(100 + 99 + 98 + \ldots + 51) - (50 + 49 + 48 + \ldots + 1) =$
    A) 250   B) 2500   C) 2525   D) 2550

The end of the contest 5

Visit our Web site at http://www.mathleague.com

Solutions on Page 95 • Answers on Page 143

FIFTH GRADE MATHEMATICS CONTEST

Math League Press, P.O. Box 17, Tenafly, New Jersey 07670-0017

2007-08 Annual 5th Grade Contest

Spring, 2008

5

Instructions

- **Time** Do *not* open this booklet until you are told by your teacher to begin. You will have only *30 minutes* working time for this contest. You might be *unable* to finish all 30 questions in the time allowed.

- **Scores** Please remember that *this is a contest, not a test*—and there is no "passing" or "failing" score. Few students score as high as 24 points (80% correct). Students with half that, 12 points, *should be commended!*

- **Format and Point Value** This is a multiple-choice contest. Each answer is an A, B, C, or D. Write each answer in the *Answer Column* to the right of each question. A correct answer is worth 1 point. Unanswered questions get no credit. You **may** use a calculator.

Copyright © 2008 by Mathematics Leagues Inc.

2007-2008 5TH GRADE CONTEST

1. $110 - 10 + 120 - 20 + 130 - 30 =$
   A) 100 B) 200 C) 300 D) 400

2. Each roller coaster ride costs 3 tickets. How many more such rides can I take with 30 tickets than I can with 18 tickets?
   A) 4 B) 6 C) 10 D) 12

3. $10 \times 10 \times 11 \times 11 \times 0 =$
   A) 0 B) 42 C) 220 D) 12 100

4. Of my 48 fish, 1 of every 3 is a Bigfish. I have _?_ Bigfish.
   A) 16 B) 24
   C) 32 D) 36

5. Which is 10° colder than 45°?
   A) 34° B) 35° C) 54° D) 55°

6. 1 week + 20 days = 3 weeks + _?_ days
   A) 0 B) 6 C) 7 D) 13

7. Each of my play's 3 acts has 4 10-minute scenes. My play lasts
   A) 43 minutes B) 1 hr. 10 mins. C) 1 hr. 20 mins. D) 2 hours

8. An octagon has twice as many sides as a
   A) triangle B) square
   C) pentagon D) hexagon

9. Which product is an odd number?
   A) 23 × 24 B) 23 × 25
   C) 24 × 25 D) 24 × 26

10. 989 is 98 more than
    A) 881 B) 887 C) 891 D) 898

11. Increasing the numbers 3, 9, 27, 81 by 3 each increases their sum by
    A) 3 B) 4 C) 9 D) 12

Go on to the next page ⏵ **5**

2007-2008 5TH GRADE CONTEST

12. Of the following whole numbers, which is *not* divisible by 3?
    A) 936   B) 957   C) 645   D) 629

13. Together we have $72. I have three times as much as you. I have
    A) $54   B) $48   C) $24   D) $18

14. After a dog took its football 3 steps east, 4 steps south, 5 steps west, 2 steps east, and 6 steps north, the dog was _?_ steps north of its starting point.
    A) 2   B) 3   C) 4   D) 5

15. The average of 2 twos and 6 sixes is
    A) 1   B) 4 fours   C) 5   D) 8

16. The smallest multiple of 7 that's greater than 2007 is
    A) 2008   B) 2009   C) 2014   D) 2016

17. Of 60 flamingos, twice as many stood on 2 legs as stood on 1. All together, on how many legs did all of these flamingos stand?
    A) 80   B) 90   C) 100   D) 120

18. Of the following quotients, which has the greatest remainder?
    A) 8888 ÷ 5   B) 7777 ÷ 5   C) 6666 ÷ 5   D) 4444 ÷ 5

19. $8 in dimes is the same number of coins as _?_ in quarters.
    A) $8   B) $20   C) $25   D) $40

20. Add the odd numbers between 6 and 12. How much greater is their sum than the sum of the even numbers between 5 and 11?
    A) 3   B) 4   C) 5   D) 6

21. Al's age + Bo's age + Cal's age = 48. Six years ago, that sum was
    A) 24   B) 30   C) 36   D) 42

22. If the value of my Endrun stock, now worth $128, drops by half each day, its value at the end of the 4th day will be
    A) $32   B) $16   C) $8   D) $4

Go on to the next page ⟶ **5**

2007-2008 5TH GRADE CONTEST

23. Which pair has the same greatest common factor as 36 & 63?
    A) 15 & 51   B) 24 & 42   C) 35 & 53   D) 45 & 54

24. A square of perimeter 36 is divided into 9 smaller squares. The perimeter of the shaded region is 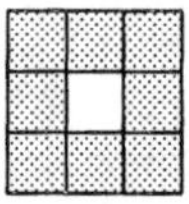
    A) 40   B) 45   C) 48   D) 96

25. If the difference between two basketball scores is 2, then the ones' digit of their sum could *never* be
    A) 0   B) 1   C) 2   D) 4

26. How many minutes before 5 P.M. is 500 minutes after 5 A.M. the same day?
    A) 240   B) 220   C) 180   D) 100

27. The average of 48 different numbers is 64. One of these numbers is 64. What is the sum of the other numbers?
    A) 47×64 B) 48×63 C) 47×63 D) 63×64

28. I counted to 600, first by 4's and then by 6's. Of the numbers that I called out the 1st time, how many did I also call out the 2nd time?
    A) 1   B) 12
    C) 25   D) 50

29. If 8 sips = 3 gulps, and 5 gulps = 2 guzzles, then _?_ sips = 24 guzzles.
    A) 24   B) 72   C) 160   D) 240

30. The figure △ contains 1 big and 2 small triangles. How many different triangles of all sizes are shown at the right? 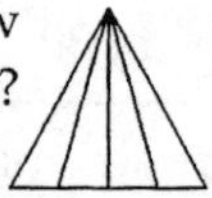
    A) 5   B) 6   C) 8   D) 10

The end of the contest 5

Visit our Web site at http://www.mathleague.com

Solutions on Page 99 • Answers on Page 144

FIFTH GRADE MATHEMATICS CONTEST

Math League Press, P.O. Box 17, Tenafly, New Jersey 07670-0017

2008-2009 Annual 5th Grade Contest

Spring, 2009

Instructions

5

- **Time** Do *not* open this booklet until you are told by your teacher to begin. You will have only *30 minutes* working time for this contest. You might be *unable* to finish all 30 questions in the time allowed.

- **Scores** Please remember that *this is a contest, and not a test*—there is no "passing" or "failing" score. Few students score as high as 24 points (80% correct). Students with half that, 12 points, *should be commended!*

- **Format and Point Value** This is a multiple-choice contest. Each answer is an A, B, C, or D. Write each answer in the *Answer Column* to the right of each question. A correct answer is worth 1 point. Unanswered questions receive no credit. You **may** use a calculator.

Copyright © 2009 by Mathematics Leagues Inc.

2008-2009 5TH GRADE CONTEST

1. 4 + 9 + 14 = 2 + 7 + 12 + _?_

   A) 2   B) 3   C) 6   D) 7

2. On Monday, Ivan's Ice Cream Stand sold only 6 cherry, 5 vanilla, 3 lime, and 2 chocolate cones. The number of cherry cones sold was _?_ less than the total number of other cones sold.

   A) 1   B) 2   C) 3   D) 4

3. How many degrees does one angle of a square have?

   A) 45   B) 60   C) 90   D) 100

4. Which of the following has the smallest value?

   A) 60 × 30   B) 6 × 3000   C) 300 × 60   D) 3 × 6000

5. (9 × 6 × 3) ÷ (3 × 2 × 1) =

   A) 3   B) 6   C) 9   D) 27

6. The ones' digit of 24 × 46 × 68 is

   A) 2   B) 4   C) 6   D) 8

7. 3 dozen = _?_ pairs

   A) 36   B) 18   C) 12   D) 6

8. Barry the bullfrog weighs four times as much as Taylor the tree frog. If Barry weighs 200 g, then Taylor weighs _?_ g.

   A) 40   B) 50   C) 250   D) 800

9. Which of the following is *not* a quadrilateral?

   A) pentagon   B) rectangle   C) square   D) parallelogram

10. 13 dimes and 13 pennies is equal to _?_ cents.

    A) 130   B) 133   C) 143   D) 146

11. If the sum of two whole numbers is 16 and their difference is 4, then their product is

    A) 56   B) 60   C) 64   D) 78

12. One full day + 30 hours = three full days − _?_ hours

    A) 6   B) 12   C) 18   D) 24

Go on to the next page))) **5**

2008-2009 5TH GRADE CONTEST

13. Which of the following quotients is greater than 4?

    A) 240 ÷ 56   B) 168 ÷ 42   C) 224 ÷ 58   D) 138 ÷ 38

14. If Edwin bikes 3 km every 10 minutes, how far does he bike in 2 hours?

    A) 9 km   B) 18 km   C) 24 km   D) 36 km

15. Which of the following is twice the thousands' digit of 12 345?

    A) 2   B) 4   C) 6   D) 8

16. 4 × 4 × 4 × 4 × 4 × 4 =

    A) 16 × 16 × 16   B) 16 + 16 + 16   C) 16 × 16   D) 16 + 16

17. Juan's age is a prime number. His age could be

    A) 42   B) 43   C) 49   D) 51

18. The perimeter of an equilateral triangle is 24. What is the length of one of its sides?

    A) 4   B) 6   C) 8   D) 12

19. Three days after tomorrow is how many days after the day before yesterday?

    A) 4   B) 5   C) 6   D) 7

20. The number of non-zero digits in the product of 4000 and 3000 is

    A) 1   B) 2   C) 6   D) 8

21. Frank stands in line and has 76 people behind him. If there are a total of 110 people in line, how many are in front of Frank?

    A) 33   B) 34   C) 35   D) 36

22. Gina buys her pens at a store that charges 80 cents for the first pen and 65 cents for each additional pen. How much would she pay to purchase 7 pens?

    A) $4.25   B) $4.40   C) $4.55   D) $4.70

23. If 5 blorks weigh as much as 9 glorks, then 90 glorks weigh as much as _?_ blorks.

    A) 10   B) 18   C) 45   D) 50

Go on to the next page ⟶ 5

24. A line is drawn through the center of a rectangle, dividing the rectangle into two squares. If the original rectangle has a length of 6 and a width of 3, what is the perimeter of one of the squares?

    A) 3   B) 6   C) 9   D) 12

25. $10 \times 30 \times 50 = 15 \times 25 \times \underline{\ ?\ }$

    A) 35   B) 40   C) 45   D) 50

26. On each day after the first day of a three-day festival, the attendance is lower than it was on the day before. Each time the attendance decreases, it decreases by the same number of people. If 600 000 people attend the festival on the second day, what is the average daily attendance at this three-day festival?

    A) 200 000   B) 300 000   C) 600 000   D) 1 200 000

27. How many 3-digit whole numbers are there such that the hundreds' digit is equal to the sum of the tens' digit and the ones' digit?

    A) 18   B) 20   C) 36   D) 54

28. Isabella has only quarters, dimes, nickels, and pennies. She has a different number of each type of coin. If Isabella's 10 coins have a total value of 83¢, she has more _?_ than any other type of coin.

    A) pennies   B) nickels   C) dimes   D) quarters

29. Jackie has been the first student to arrive at school 5 out of the last 8 school days. At this rate, she'd be the first to arrive on _?_ of the next 40 school days.

    A) 8   B) 13   C) 20   D) 25

30. Jake has 100 square tiles, each of which has a perimeter of 4 m. If he forms a rectangle using all of the tiles, what is the smallest perimeter the rectangle he forms can have?

    A) 40 m   B) 100 m   C) 202 m   D) 404 m

The end of the contest 📖 **5**

Visit our Web site at http://www.mathleague.com
Solutions on Page 103 • Answers on Page 145

FIFTH GRADE MATHEMATICS CONTEST

Math League Press, P.O. Box 17, Tenafly, New Jersey 07670-0017

2009-2010 Annual 5th Grade Contest

Spring, 2010

Instructions

5

- **Time** Do *not* open this booklet until you are told by your teacher to begin. You will have only *30 minutes* working time for this contest. You might be *unable* to finish all 30 questions in the time allowed.

- **Scores** Please remember that *this is a contest, and not a test*—there is no "passing" or "failing" score. Few students score as high as 24 points (80% correct). Students with half that, 12 points, *should be commended!*

- **Format and Point Value** This is a multiple-choice contest. Each answer is an A, B, C, or D. Write each answer in the *Answer Column* to the right of each question. A correct answer is worth 1 point. Unanswered questions receive no credit. You **may** use a calculator.

Copyright © 2010 by Mathematics Leagues Inc.

2009-2010 5TH GRADE CONTEST

1. Ira is a tuba player. There are 9 tuba players ahead of Ira in line, and 20 behind him. In all, how many tuba players are in the line?
   A) 28 B) 29 C) 30 D) 31

2. $(2 \times 3 \times 4 \times 5) \div 12 =$
   A) 6 B) 8 C) 10 D) 12

3. Rachel wants to cut a single 30 m long piece of rope into smaller pieces. She can cut at most _?_ pieces that are 4 m long.
   A) 6 B) 7 C) 8 D) 9

4. Sam finds 4 shells each minute at the beach. How many shells does she find in an hour?
   A) 15 B) 20 C) 64 D) 240

5. $(32 \div 4) + (48 \div 6) + (64 \div 8) = 8 \times$ _?_
   A) 3 B) 6 C) 8 D) 12

6. 9×9 is equal to
   A) $999 \div 9$ B) $90 + 9$ C) $90 - 9$ D) 99

7. Tom, Dick, and Harry go to the movies together. Tom pays for all 3 tickets with a $50 bill. If he gets $18.50 in change and each ticket costs the same amount, what is the cost of each ticket?
   A) $10.50 B) $11.50 C) $12.50 D) $18.50

8. Of the following numbers, which is divisible by 3?
   A) 1050 B) 2024 C) 3058 D) 4022

9. If 4 of Tom's friends can paint his fence in 12 hours, and all of his friends work at the same rate, then how many hours would it take for 8 of his friends to paint his fence?
   A) 4 B) 6 C) 8 D) 24

10. 2:50 P.M. is _?_ minutes before 3:15 P.M.
    A) 25 B) 45 C) 65 D) 75

11. The ones digit of the product
    $2009 \times 2008 \times 2007 \times 2006 \times 2005 \times 2004 \times 2003 \times 2002$ is
    A) 8 B) 6 C) 4 D) 0

Go on to the next page ⟫ **5**

	Answers
12. The total value of 25 quarters, 10 dimes, 5 nickels, and 1 penny is A) $7.01 B) $7.16 C) $7.51 D) $8.11	12.
13. The largest multiple of 8 that is less than 300 is A) 298 B) 296 C) 294 D) 292	13.
14. Of the following products, which is 3 less than (2010 ÷ 30)? A) 7 × 10 B) 6 × 12 C) 5 × 14 D) 4 × 16	14.
15. Rose asks 180 dentists whether she should chew sugarless gum. If 4 out of 5 of them say she should, how many do NOT say she should? A) 9 B) 36 C) 72 D) 144	15.
16. 20 hundreds + 20 tens + 20 ones = A) 222 B) 2022 C) 2040 D) 2220	16.
17. When 161 616 169 is divided by 8, the remainder is A) 1 B) 3 C) 5 D) 7	17.
18. Each of the 300 5th graders at Einstein School got a score of 10 on a science contest. Each of the 300 students at Copernicus School scored a 6 on this same contest. What was the average score of all 600 students? A) 10 B) 8 C) 7 D) 6	18.
19. The time 7 hours before 7 P.M. is the same as the time 7 hours after A) 5 A.M. B) 7 A.M. C) 2 P.M. D) 7 P.M.	19.
20. If 1 blorp = 4 blurps, and 2 blurps = 5 blaps, then 6 blorps = _?_ blaps. A) 16 B) 24 C) 48 D) 60	20.
21. Gavin is a monkey and Hal is his owner. Hal's age is 7 years greater than 3 times Gavin's age. If Gavin is 11, what is the difference between Gavin's age and Hal's age? A) 15 B) 18 C) 29 D) 51	21.
22. The sum of two consecutive whole numbers is 4019. What is the difference between those two numbers? A) 1 B) 9 C) 2009 D) 2010	22.

Go on to the next page ⟫▶5

2009-2010 5TH GRADE CONTEST

23. Of the following polygons, all have the same number of sides *except* a
    A) hexagon   B) rhombus   C) trapezoid   D) parallelogram

24. Alice and Bob each write 10 letters. Charlie, Dan, and Evelyn each write 20 letters. What is the average number of letters these five people write?
    A) 18   B) 16   C) 15   D) 12

25. There are exactly 3 prime numbers between
    A) 0 and 6   B) 6 and 12
    C) 12 and 18   D) 18 and 24

26. Jesse and James each had $10 to spend on school supplies. Jesse spent his $10 on pens that cost $0.50 each. James spent his $10 on pencils that cost $0.20 each. The difference between the number of pencils that James bought and the number of pens that Jesse bought is
    A) 10   B) 25   C) 30   D) 50

27. As shown, rectangle R consists of 6 squares of perimeter 8. What is the area of rectangle R?
    A) 10   B) 12   C) 20   D) 24

28. I divided 123 by 4 and got the quotient and the remainder. This quotient divided by this remainder equals
    A) 5   B) 10   C) 15   D) 23

29. Of the whole numbers less than ten thousand, how many contain each of the digits 1, 2, 3, and 4?
    A) 6   B) 12   C) 18   D) 24

30. A piece of paper with four 5-digit numbers written on it got wet. The numbers' middle digits now cannot be read. Each middle digit has been replaced with a ■. Which of the numbers is divisible by 4?
    A) 24■58   B) 53■49   C) 25■74   D) 67■92

The end of the contest 🖎 5

Visit our Web site at http://www.mathleague.com
Solutions on Page 107 • Answers on Page 146

FIFTH GRADE MATHEMATICS CONTEST

Math League Press, P.O. Box 17, Tenafly, New Jersey 07670-0017

2010-2011 Annual 5th Grade Contest

Spring, 2011

5

Instructions

- **Time** Do *not* open this booklet until you are told by your teacher to begin. You will have only *30 minutes* working time for this contest. You might be *unable* to finish all 30 questions in the time allowed.

- **Scores** Please remember that *this is a contest, and not a test*—there is no "passing" or "failing" score. Few students score as high as 24 points (80% correct). Students with half that, 12 points, *should be commended!*

- **Format and Point Value** This is a multiple-choice contest. Each answer is an A, B, C, or D. Write each answer in the *Answer Column* to the right of each question. A correct answer is worth 1 point. Unanswered questions receive no credit. You **may** use a calculator.

Copyright © 2011 by Mathematics Leagues Inc.

2010-2011 5TH GRADE CONTEST

1. 2010 + 2020 + 2030 = 2000 × 3 + _?_
   A) 30   B) 50   C) 60   D) 80

2. I have 50 fish. How many bowls do I need if I want to put 2 fish in each bowl?
   A) 25   B) 52   C) 100   D) 502

3. 10 + 8 × 6 − 4 ÷ 2 =
   A) 18   B) 27   C) 52   D) 56

4. Julie has 4 quarters, 4 dimes, 4 nickels, and 4 pennies. She has
   A) $1.64   B) $2.44   C) $3.64   D) $4.44

5. Which of the following products is a multiple of 6?
   A) 500 × 2   B) 400 × 4   C) 300 × 3   D) 200 × 8

6. A _?_ has an odd number of sides.
   A) hexagon   B) pentagon   C) rectangle   D) rhombus

7. One of my two brothers is 4 years older than the other. If the sum of their ages is 38, the older brother is _?_ years old.
   A) 17   B) 21   C) 23   D) 27

8. Lenny has a collection of giant books that are either hardcover or softcover. If Lenny has 63 books, and he has 2 hardcover books for every 7 softcover books, how many hardcover books does Lenny have?
   A) 14   B) 18   C) 43   D) 49

9. Of the following numbers, which leaves a remainder of 2 when divided by 4?
   A) 2224   B) 3350   C) 4481   D) 5523

10. Rose is 150 cm tall. Rose's older brother Quentin is 10 cm taller than Rose is. Rose's younger brother Sam is 4 cm shorter than Rose is. What is the average of the heights of Rose, Sam, and Quentin?
    A) 148 cm   B) 150 cm   C) 152 cm   D) 154 cm

11. What is the difference between the two whole numbers whose product is 36 and whose sum is 13?
    A) 8   B) 7   C) 6   D) 5

Go on to the next page))))▶ 5

2010-2011 5TH GRADE CONTEST

12. Professor Promptov insists that all of his students arrive at his class exactly 7 hours and 77 minutes before 7 P.M. That means they must arrive at

    A) 10:43 A.M. B) 11:17 A.M.
    C) 12:07 P.M. D) 12:17 P.M.

13. 60 more than 6 × _?_ is 60.

    A) 0   B) 10   C) 12   D) 20

14. The sum of three different positive even numbers could be

    A) 10   B) 13   C) 14   D) 19

15. In my suitcase I have 5 sweaters and 6 pairs of pants. If I make an outfit of a sweater and a pair of pants, how many different outfits can I select?

    A) 11   B) 22   C) 25   D) 30

16. Zoe is exactly twelve and one-half years old, and her niece Yolanda is one-third as old as Zoe. Yolanda is _?_ months old.

    A) 48   B) 50   C) 100   D) 150

17. The number of balloons the balloon man has is the smallest whole number divisible by 1, 2, 3, 4, 5 and 6. He has _?_ balloons.

    A) 21   B) 60   C) 120   D) 720

18. My teacher uses 5 pieces of chalk every 6 days. He will use _?_ pieces of chalk in 30 days.

    A) 25   B) 30   C) 35   D) 36

19. My favorite prime is 6 less than another prime. My favorite prime could be

    A) 19   B) 29   C) 35   D) 37

20. There are _?_ odd 2-digit whole numbers.

    A) 45   B) 50   C) 55   D) 60

21. I burn 4 candles each day, and need enough candles to last a full week. If candles are sold in packs of 5, I need at least _?_ packs.

    A) 4   B) 5   C) 6   D) 7

Go on to the next page))))▶ 5

2010-2011 5TH GRADE CONTEST

		Answers
22.	Each of 6 circus elephants weighs at least 3500 kg. The total weight of all 6 is 24 000 kg. The *most* any one of these elephants can possibly weigh is _?_ kg. A) 4000 B) 5000 C) 5500 D) 6500	22.
23.	Speedy drives from 10 A.M. until 11:30 A.M. at a rate of 120 km per hour. Speedy drives _?_ km. A) 120 B) 150 C) 180 D) 210	23.
24.	I multiply all the even numbers between 50 and 90. The ones digit of this product is A) 0 B) 2 C) 4 D) 8	24.
25.	Each of two friends chooses a whole number from 1 through 5. I multiply the two numbers they choose. There are _?_ different possible values of the product. A) 14 B) 17 C) 20 D) 25	25.
26.	I have 7 coins with a total value of 77¢. If my largest coin is a quarter, I have _?_ dimes. A) 1 B) 2 C) 3 D) 4	26.
27.	I am standing 4 km from my house and 7 km from my school. The distance between my house and my school *cannot* be A) 12 km B) 7 km C) 6 km D) 3 km	27.
28.	The product of 6 whole numbers, not all different, is 60. Their sum could *not* be A) 14 B) 15 C) 16 D) 17	28.
29.	At the fruit stand, 3 apples and 2 oranges cost $3.20, and 2 apples and 3 oranges cost $2.80. What is the cost of 1 apple and 1 orange? A) $1.00 B) $1.10 C) $1.15 D) $1.20	29.
30.	Brooke's empty tub fills in 20 minutes with the drain plugged, and her full tub drains in 10 minutes with the water off. How many minutes would it take the full tub to drain while the water is on? A) 12 B) 15 C) 20 D) 30	30.

The end of the contest ✍ 5

Visit our Web site at http://www.mathleague.com
Solutions on Page 111 • Answers on Page 147

6th Grade Contests

2006-2007 through 2010-2011

SIXTH GRADE MATHEMATICS CONTEST

Math League Press, P.O. Box 17, Tenafly, New Jersey 07670-0017

2006-2007 Annual 6th Grade Contest

Tuesday, February 20 or 27, 2007

Instructions

6

- **Time** You will have only *30 minutes* working time for this contest. You might be *unable* to finish all 40 questions in the time allowed.

- **Scores** Please remember that *this is a contest, not a test*—and there is no "passing" or "failing" score. Few students score as high as 30 points (75% correct). Students with half that, 15 points, *should be commended!*

- **Format and Point Value** This is a multiple-choice contest. Each answer is an A, B, C, or D. Write each answer in the *Answers* column to the right of each question. A correct answer is worth 1 point. Unanswered questions get no credit. You **may** use a calculator.

Copyright © 2007 by Mathematics Leagues Inc.

2006-2007 6TH GRADE CONTEST

#	Question	Answers
1.	How many dozens are in 24 + 36 + 48 + 60? A) 7 B) 9 C) 10 D) 14	1. d
2.	If gumballs cost a quarter each, then the cost of the 40 gumballs that I ate today was A) $100 B) $65 C) $10 D) $1	2. c
3.	$45 \div 3 = 3 \times \underline{?}$ A) 3 B) 5 C) 9 D) 15	3. b
4.	$1.25125 \times \frac{5}{4} = 1.25125 \times \underline{?}$ A) 1.20 B) 1.25 C) 1.40 D) 1.75	4. b
5.	If the area of each of the 3 squares shown is 9, what is the perimeter of the entire figure? A) 18 B) 24 C) 27 D) 36	5. b
6.	Of the following quotients, which is greatest? A) $64 \div 4$ B) $112 \div 7$ C) $144 \div 9$ D) $194 \div 12$	6. d
7.	If twice my height is 450 cm, then my height divided by 5 is A) 180 cm B) 135 cm C) 90 cm D) 45 cm	7. d
8.	The largest power of 20 that's a factor of $20 \times 40 \times 60$ is A) 20^1 B) 20^3 C) 20^6 D) 20^{2400}	8. b
9.	If we eat 3 slices every 20 minutes, then we'll eat $\underline{?}$ slices in 1 hour. A) 9 B) 12 C) 15 D) 45	9. a
10.	$99 \times 99 = \underline{?} - 99$ A) 199×99 B) 198×99 C) 100×100 D) 100×99	10. d
11.	I can write the number $\underline{?}$ as the product of 3 numbers, all primes. A) 25 B) 35 C) 45 D) 55	11. c
12.	200% of 6 = 300% of $\underline{?}$ A) 2 B) 3 C) 4 D) 12	12. c
13.	Of 132 clowns, half are redheads. Half the rest are blonde. The ratio of the number of blonde clowns to the number of redheaded clowns is A) 33:66 B) 44:66 C) 66:66 D) 55:77	13. a
14.	The number midway between 2 and 12 is A) 5 B) 6 C) 7 D) 8	14. c
15.	$3 \times 10^4 + 4 \times 10^3 + 5 \times 10^2 + 6 \times 10^2 + 7 \times 10 =$ A) 345 670 B) 35 170 C) 34 170 D) 34 567	15. b

Go on to the next page ▶ **6**

#	Question	Answer
16.	To get the greatest possible result, round 4454 to the nearest A) one B) ten C) hundred D) thousand	16. C
17.	An eclipse lasting from 3:45 P.M. to 6:35 P.M. was half over at A) 4:50 P.M. B) 4:55 P.M. C) 5:05 P.M. D) 5:10 P.M.	17. d
18.	(hundreds' digit of 24 683 579) × (ten-thousands' digit of 24 683 579) = A) 56 B) 40 C) 30 D) 10	18. b
19.	The average price of twelve $4 tickets and four $12 tickets is A) $4 B) $6 C) $8 D) $12	19. b
20.	2×4×6 is a factor of A) 2×3×4×5 B) 3×4×5×6 C) 4×5×6×7 D) 5×6×7×8	20. d
21.	On a calendar, I put 1 grain of sand on May 1, 2 grains on May 2, 4 grains on May 3, and so forth, doubling the number of grains each day. On what day will I put down the 500th grain? A) May 9 B) May 10 C) May 25 D) June 19	21. a
22.	Divide the remainder in (888 ÷ 77) by 6. The new remainder is A) 0 B) 1 C) 3 D) 5	22. d
23.	The total value of an equal number of pennies and dimes *can't* be A) $10.01 B) $11.00 C) $11.11 D) $11.12	23. d
24.	$100\,001^2$ exceeds $100\,000^2$ by A) 200 001 B) 100 001 C) $200\,001 \times 10^6$ D) $100\,001 \times 10^6$	24. a
25.	I wrote two whole numbers whose sum is less than their product. What is the smallest possible whole number I wrote? A) 0 B) 1 C) 2 D) 3	25. C
26.	A circular pool that's 2 m wide has a circumference of A) π m B) 2π m C) 4π m D) 5π m	26. b
27.	$\sqrt{40 \times 90}$ = A) 36 B) 50 C) 60 D) 120	27.
28.	The only prime factor of ? is 2. A) 2222 B) 2468 C) 4848 D) 8192	28. d
29.	The l.c.m. of 8 and 10 equals the greatest common factor of 80 and A) 160 B) 120 C) 100 D) 80	29. b

Go on to the next page ⮞ **6**

#	Question	Answers
30.	At most ? squares with side-length 4 can fit inside a rectangle with area 2000 without overlapping. A) 121 B) 125 C) 250 D) 500	30. B
31.	What percent of a 3-hour concert will I see if I'm there for 18 minutes? A) 6% B) 10% C) 21% D) 30%	31. B
32.	If an equilateral triangle has integer sides, its perimeter *cannot* be A) 915 B) 615 C) 315 D) 115	32. D
33.	How many different positive integers are factors of $2006 = 2 \times 17 \times 59$? A) 4 B) 6 C) 8 D) 9	33. C
34.	Of 60 kids, if 20 like math, 30 like art, and 10 like both, then how many like neither? A) 0 B) 10 C) 20 D) 30	34. C
35.	If $1 + 3 + 5 + \ldots + 99 = 2500$, then $2 + 4 + 6 + \ldots + 100 =$ A) 2550 B) 2599 C) 2600 D) 5000	35. A
36.	Bob counted by 3's starting with 2. He got 2, 5, 8, 11, 14, 17, ... Ann counted by 2's starting with 3. She got 3, 5, 7, 9, 11, 13, ... Bob's 100th number is ? more than Ann's 100th number. A) 97 B) 98 C) 99 D) 100	36. B
37.	A tree with 20 times as many leaves as branches could have A) 399 leaves B) 400 leaves C) 401 leaves D) 410 leaves	37. B
38.	After I cut some red tape into 50 pieces, I then cut each of the 50 pieces into 5 smaller pieces. *At most* how many cuts did I make altogether? A) 249 B) 250 C) 299 D) 300	38. A
39.	When folded along the lines shown, which *cannot* form a cube? A) B) C) D)	39. A
40.	If 4 flips = 3 flops, and 2 flops = 4 flaps, then ? flips = 6 flaps. A) 2 B) 3 C) 4 D) 6	40. C

The end of the contest 6

SIXTH GRADE MATHEMATICS CONTEST

Math League Press, P.O. Box 17, Tenafly, New Jersey 07670-0017

2007-2008 Annual 6th Grade Contest

Tuesday, February 19 or 26, 2008

Instructions

- **Time** You will have only *30 minutes* working time for this contest. You might be *unable* to finish all 40 questions in the time allowed.
- **Scores** Please remember that *this is a contest, not a test*—and there is no "passing" or "failing" score. Few students score as high as 30 points (75% correct). Students with half that, 15 points, *should be commended!*
- **Format and Point Value** This is a multiple-choice contest. Each answer is an A, B, C, or D. Write each answer in the *Answers* column to the right of each question. A correct answer is worth 1 point. Unanswered questions get no credit. You **may** use a calculator.

Copyright © 2008 by Mathematics Leagues Inc.

2007-2008 6TH GRADE CONTEST

		Answers
1.	Sue is twice as old as Sam. If Sam's age is 18, then Sue's age is A) 9 B) 27 C) 32 D) 36	1. D
2.	In the "Couples Only" Run, 4 dozen people participated. How many couples participated in the "Couples Only" Run? A) 12 B) 24 C) 48 D) 96	2. B
3.	The Run started at 9:15 A.M. If it ended at 4:11 P.M., then the Run was half over at ? P.M. A) 12:43 B) 1:13 C) 1:33 D) 3:28	3. A
4.	Which is 1 less than a prime factor of 375? A) 4 B) 24 C) 124 D) 374	4. A
5.	If today is Tuesday, what day was it 43 days ago? A) Sunday B) Monday C) Wednesday D) Thursday	5. B
6.	If a triangle has perimeter 72, the average length of a side is A) 3 B) 18 C) 24 D) 36	6. C
7.	In which of the following is the remainder an odd number? A) 9898 ÷ 3 B) 9898 ÷ 4 C) 9898 ÷ 6 D) 9898 ÷ 7	7. A
8.	(20+30+40+50) ÷ 4 = (10+15+20+25) ÷ ? A) 1 B) 2 C) 4 D) 8	8. B
9.	In 12 345 678 910, the ratio (# of odd digits) : (# of even digits) = A) 1:2 B) 2:3 C) 1:1 D) 6:5	9. D
10.	Each of the following is a factor of 23 × 24 × 25 × 26 *except* A) 4 B) 5 C) 6 D) 7	10. D
11.	Salty Sandy, who began with $1, spent 45¢ on a jar of salt. After Peppy Patty, who began with $1.45, bought a jar of pepper for ?, she and Salty Sandy were left with the same amount of money. A) 45¢ B) 55¢ C) 90¢ D) $1	11. C
12.	$3^2 \times 6^2 \times 9^2$ has the same value as A) $3^6 \times 6^3$ B) $3^6 \times 9^3$ C) $6^2 \times 9^3$ D) $6^3 \times 9^2$	12. C
13.	I can cut a square of area 36 into at most ? squares of perimeter 4. A) 4 B) 9 C) 18 D) 36	13. D
14.	Of the following, which ratio *most nearly* equals the ratio 3:2? A) 13:12 B) 15:12 C) 19:12 D) 23:12	14. C
15.	Find the average of one 1, two 2s, three 3s, and four 4s. A) 2 B) 3 C) 4 D) 5	15. B

Go on to the next page

#	Question	Answer
16.	Ed began by taking a sip of milk. He took one more sip after every 2 bites of his sandwich. Ed took 21 bites of his sandwich, so he took _?_ sips of milk. A) 7 B) 8 C) 11 D) 12	16. C
17.	$444 + 888 = 333 \times$ _?_ A) 2 B) 3 C) 4 D) 6	17. C
18.	$123\,123\,123\,123 \div 123\,123 =$ A) 1 000 001 B) 1 001 001 C) 1 010 101 D) 1 111 111	18. a
19.	The product of all the factors of 12 is A) 12 B) 12×12 C) $6 \times 8 \times 12$ D) $12 \times 12 \times 12$	19. d
20.	In 2 years, I'll be twice my age 4 years ago. How old am I now? A) 8 B) 10 C) 12 D) 14	20. b
21.	A rectangle with perimeter 20 has an area of at most A) 20 B) 25 C) 100 D) 400	21. b
22.	Altogether, an octagon, a hexagon, and a pentagon have _?_ sides. A) 18 B) 19 C) 20 D) 21	22. b
23.	The greatest common divisor of $3 \times 4 \times 5$ and $13 \times 14 \times 15$ is A) $3 \times 4 \times 5$ B) $2 \times 3 \times 5$ C) 3×5 D) 5	23. b
24.	The product of the numbers 3 and _?_ is equal to their sum. A) $\frac{2}{3}$ B) $\frac{3}{4}$ C) $\frac{4}{3}$ D) $\frac{3}{2}$	24. d
25.	$99\,990^2 \div 9999^2 =$ A) 1 B) 10 C) 100 D) 1000	25. C
26.	If the center of a circular pool is 2 m from the pool wall, then the circumference of the pool is _?_ m. A) π B) 2π C) 4π D) 8π	26. C
27.	If my daughter hoses me once every 4 sunny days, and if it's sunny 2 days out of 3, then I expect to get hosed _?_ times in 36 days. A) 6 B) 9 C) 12 D) 24	27. c
28.	$50 \times 40 \times 30 \times 20 \times 10 = 25 \times 20 \times 15 \times 10 \times 5 \times$ _?_ A) 2 B) 10 C) 20 D) 32	28. d
29.	100% of $10^2 = 10 \times$ _?_ A) 10 B) 10^2 C) 10^3 D) 10^4	29. a

#	Question	Answer
30.	How many of the numbers 3, 12, 24, and 48 are factors of every whole number that is divisible by both 6 and 8? A) 1 B) 2 C) 3 D) 4	30. C
31.	How long is a radius of the largest circle whose points are all either inside of or on a square whose perimeter is 8? A) 8 B) 4 C) 2 D) 1	31. d
32.	I plan to buy 2 more $6 puzzles than $5 puzzles, so for exactly $100, I plan to buy ? puzzles. A) 17 B) 18 C) 19 D) 20	32. b
33.	$\sqrt{3 \times 12} \times \sqrt{4 \times 9} =$ A) 2×18 B) 18×18 C) 3×2 D) 36×36	33. a
34.	If I sit in a row of 26 seats, then the ratio of the number of seats on my left to the number of seats on my right could be A) 1:1 B) 1:2 C) 1:3 D) 1:4	34. d
35.	The ones' digit of the product of 2008 consecutive integers, each greater than 0, can have up to ? different possible values. A) 1 B) 5 C) 9 D) 10	35. a
36.	If $2+4+6+\ldots+100 = 2550$, then $1+3+5+\ldots+99 =$ A) 2500 B) 2475 C) 2450 D) 1275	36. a
37.	With his hat on, the Invisible Man is 180 cm tall. His hat adds 20% to his height. Without his hat, the Invisible Man's height is ? cm. A) 144 B) 150 C) 160 D) 216	37. a
38.	For ? of the 1440 minutes in a 24-hour period, one or more 5s appear in a digital clock's display. A) 362 B) 450 C) 472 D) 492	38. b
39.	A square whose area is 64 is split into two rectangles whose areas differ by 16. The perimeter of the smaller rectangle is A) 20 B) 22 C) 24 D) 26	39. b
40.	How many different triangles of all sizes can be traced along lines already drawn in the diagram? A) 16 B) 15 C) 13 D) 11	40. a

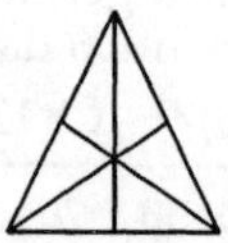

The end of the contest **6**

Visit our Web site at http://www.mathleague.com

Solutions on Page 121 • Answers on Page 149

SIXTH GRADE MATHEMATICS CONTEST

Math League Press, P.O. Box 17, Tenafly, New Jersey 07670-0017

2008-2009 Annual 6th Grade Contest

Tuesday, February 17 or 24, 2009

6

Instructions

- **Time** You will have only *30 minutes* working time for this contest. You might be *unable* to finish all 40 questions in the time allowed.

- **Scores** Please remember that *this is a contest, not a test*—and there is no "passing" or "failing" score. Few students score as high as 30 points (75% correct). Students with half that, 15 points, *should be commended!*

- **Format and Point Value** This is a multiple-choice contest. Each answer is an A, B, C, or D. Write each answer in the *Answers* column to the right of each question. A correct answer is worth 1 point. Unanswered questions get no credit. You **may** use a calculator.

Copyright © 2009 by Mathematics Leagues Inc.

2008-2009 6TH GRADE CONTEST

#	Question	Answer
1.	$25 + 35 + 45 = 60 + \underline{\ ?\ }$ A) 25 B) 35 C) 45 D) 55	1. C
2.	The number of weeks in 139 days is most nearly equal to A) 5 B) 14 C) 19 D) 20	2. D
3.	The number of days in July plus the number in August is twice the number of days in A) March B) April C) June D) November	3. A
4.	$(3 \times 1) + (3 \times 2) + (3 \times 3) + (3 \times 4) = 3 \times \underline{\ ?\ }$ A) 5 B) 1+2+3+4 C) 12 D) $1 \times 2 \times 3 \times 4$	4. B
5.	$8002 - 2008 = \underline{\ ?\ } - 2009$ A) 9003 B) 9002 C) 8003 D) 8002	5. C
6.	If both square S and equilateral triangle T have a perimeter of 60 cm, then each side of T is $\underline{\ ?\ }$ longer than each side of S. A) 3 cm B) 5 cm C) 8 cm D) 15 cm	6. B
7.	A regular polygon with perimeter 60 *cannot* have a side of length A) 30 B) 20 C) 15 D) 12	7. A
8.	The average value of \$2, \$4, \$6, \$8, and \$10 is $\underline{\ ?\ }$ pennies. A) 3000 B) 600 C) 550 D) 500	8. B
9.	The greatest odd factor of the product $1 \times 2 \times 3 \times 4 \times 5 \times 6$ is A) 5 B) 15 C) 45 D) 75	9. C
10.	50% of 30% = 15% of A) 10% B) 25% C) 100% D) 150%	10. C
11.	Before I began snacking, there were $60 \div 4 + 1 \times 3$ gumballs here. If I ate all of them, how many gumballs did I eat? A) 4 B) 18 C) 36 D) 48	11. B
12.	If 20×30 is divided by 40, the remainder is A) 30 B) 15 C) 10 D) 0	12. D
13.	$180 \div 6 = 6 \times \underline{\ ?\ }$ A) 180 B) 36 C) 30 D) 5	13. D
14.	Ann sleeps just 8 hrs. each day, so in 10 days, she's *awake* $\underline{\ ?\ }$ hrs. A) 10×16 B) 8×10 C) 8×24 D) 16×24	14. A
15.	The total value of 25 dimes is 125 times the total value of A) 1 penny B) 1 nickel C) 2 pennies D) 2 nickels	15. C
16.	(the number of cm in 1 m) : (the number of m in 1 km) = A) 100:1000 B) 1000:100 C) 1:100 D) 100:1	16. A

Go on to the next page ⟶ 6

		Answers
17.	Al ran twice as far as Bob ran. They ran a total of 18 km. How far did Al run? A) 3 km B) 6 km C) 9 km D) 12 km	17. d
18.	Since $7^2 = 49$, its ones' digit is a 9. What is the ones' digit of 2^{50}? A) 0 B) 2 C) 4 D) 8	18. c
19.	If a turkey facing north turns 225° clockwise, it will then face A) southwest B) southeast C) northwest D) northeast	19. a
20.	If I ? the number of dozens by ? , then I'll get the number of pairs. A) multiply, 6 B) divide, 6 C) multiply, 2 D) divide, 2	20. a
21.	The lengths of the three sides of a triangle could *not* be A) 1, 1, 3 B) 2, 2, 3 C) 3, 3, 3 D) 4, 4, 3	21. a
22.	A whole number divisible by 6 and by 14 need *not* be divisible by A) 21 B) 12 C) 7 D) 3	22. b
23.	Round 398° C to the nearest 10°. A) 380° C B) 390° C C) 399° C D) 400° C	23. d
24.	A painting, *Cat Smile*, is priced at $1200. Its price is increased by 10%. Its new price is then decreased by 10%. What is the final price of *Cat Smile*? A) $1212 B) $1200 C) $1188 D) $1100	24. c
25.	$10^5 + 10^6 = 10^5 \times$? A) 11 B) 12 C) 10^2 D) 10^6	25. a
26.	If 6 hoots = 3 hollers, then 10 hollers = ? hoots. A) 5 B) 13 C) 18 D) 20	26. d
27.	The product of two whole numbers is 42. Their sum *cannot* be A) 43 B) 33 C) 23 D) 13	27. b
28.	As shown, a circle of diameter 2 is drawn inside a square of side 4. To the nearest tenth, what is the perimeter of the shaded region? A) 3.4 B) 12.9 C) 22.3 D) 28.6	28. c
29.	The average of the *different* prime factors of 2009 is A) 2009 B) 147 C) 48 D) 24	29. b

Go on to the next page ⟹ **6**

		Answers
30.	Of the multiples of 7 that exceed 7, how many are factors of 700? A) 99　　B) 8　　C) 7　　D) 5	30. B
31.	If points A through I are spaced evenly on the number line, then the distance from B to H is twice the distance from A B C D E F G H I A) A to F　　B) E to G C) C to E　　D) D to G	31. D
32.	I ride my bicycle exactly 50 km every other day. In 15 weeks, I ride *at most* A) 2250 km　　B) 2600 km C) 2650 km　　D) 3000 km	32. C
33.	$\sqrt{2 \times 4 \times 8} \times \sqrt{8 \times 8} =$ A) 64　　B) 32　　C) 16　　D) 8	33. A
34.	If my school has four times as many girls as boys, then the number of girls minus the number of boys *could* be A) 2013　　B) 2011　　C) 2009　　D) 2008	34. A
35.	Of the first 100 whole numbers, ? use the digit 2 at least once. A) 20　　B) 19　　C) 11　　D) 10	35. A
36.	If I write all 26 letters of the English alphabet in alphabetical order 62 times in a row, then the 806th letter I write will be A) A　　B) E　　C) V　　D) Z	36. D
37.	If the ratio of my age now to my age 6 years ago is 3:2, then my age 4 years from now will be A) 18　　B) 20　　C) 22　　D) 24	37. C
38.	After each of 50 cards is marked with a different whole number from 1 through 50, the cards are then paired at random. *At most* how many of these 25 pairs have a sum of 25? A) 1　　B) 6　　C) 12　　D) 24	38. C
39.	The sum of the whole numbers from 1 through 100 is 5050. What is the sum of the whole numbers from 1 through 200? A) 5150　　B) 10 100　　C) 11 050　　D) 20 100	39. D
40.	In how many different ways can six identical coins be distributed among Al, Bo, and Carl so that each gets at least 1 coin? A) 10　　B) 9　　C) 8　　D) 7	40. A

The end of the contest 6

Visit our Web site at http://www.mathleague.com
Solutions on Page 125 • Answers on Page 150

SIXTH GRADE MATHEMATICS CONTEST

Math League Press, P.O. Box 17, Tenafly, New Jersey 07670-0017

2009-2010 Annual 6th Grade Contest

Tuesday, February 16 or 23, 2010

6

Instructions

- **Time** Do *not* open this booklet until you are told by your teacher to begin. You might be *unable* to finish all 40 questions in the 30 minutes allowed.
- **Scores** Please remember that *this is a contest, and not a test*—there is no "passing" or "failing" score. Few students score as high as 30 points (75% correct). Students with half that, 15 points, *should be commended!*
- **Format, Point Value, & Eligibility** Every answer is an A, B, C, or D. Write answers in the *Answers* column. A correct answer is worth 1 point. Unanswered questions get no credit. You **may** use a calculator.

Copyright © 2010 by Mathematics Leagues Inc.

2009-2010 6TH GRADE CONTEST

1. A spider has 8 legs and a tortoise has 4 legs. How many legs do 3 spiders and 3 tortoises have all together?
   A) 14 B) 17 C) 36 D) 42

2. ___?___ is divisible by 3.
   A) 2009 B) 2010 C) 2011 D) 2012

3. $4 \times 4 \times 2 \times 2 \times 4 \times 0 =$
   A) 6400 B) 64 C) 12 D) 0

4. A square has a side of length 5. What is its perimeter?
   A) 10 B) 20 C) 25 D) 50

5. $13 + (15 + 17) =$
   A) $(13 + 15) + 17$
   B) $(13 + 15) + (13 + 17)$
   C) $(13 \times 15) + (13 \times 17)$
   D) $13 \times (15 + 17)$

6. A Ferris wheel costs 50¢ per ride and a roller coaster costs $1.25 per ride. The total cost of 5 Ferris wheel rides and 10 roller coaster rides is
   A) $13 B) $14 C) $15 D) $16

7. $\frac{1}{8} + \frac{2}{8} + \frac{3}{8} =$
   A) $\frac{3}{4}$ B) $\frac{3}{8}$ C) $\frac{3}{16}$ D) $\frac{5}{24}$

8. Yesterday the train came at 8 AM, and today it came at 3 PM. How many hours passed between yesterday's and today's arrivals?
   A) 7 B) 19 C) 31 D) 35

9. $2008 + 2009 + 2010 + 2011 + 2012 =$
   A) 10050 B) 10051 C) 10052 D) 10053

10. How many prime factors does 42 have?
    A) 1 B) 2 C) 3 D) 4

11. Half the sum of the degree-measures of the angles of an isosceles triangle is
    A) 45 B) 90 C) 180 D) 360

12. Which of the following numbers is *not* the square of a whole number?
    A) 100 B) 144 C) 196 D) 200

13. The greatest common factor of 23 and 24 is
    A) 20 B) 12 C) 2 D) 1

14. $6 \times 6 \times 6 \times 6 \times 6 =$
    A) 6×5 B) 5^6 C) 6^5 D) 4^6

Go on to the next page 6

2009-2010 6TH GRADE CONTEST

15. Amy's age is three times the age of her little sister Bo. Her Uncle Charles' age is three times the sum of the ages of Amy and Bo. If Amy is 18, how old is Charles?

    A) 54 B) 60 C) 66 D) 72

16. 5 = 10% of 20% of

    A) 1000 B) 530 C) 500 D) 250

17. How many even numbers are there between 2011 and 2099?

    A) 44 B) 45 C) 88 D) 89

18. What is the average of 80, 83, 86, 89, and 92?

    A) 85 B) 85.5 C) 86 D) 86.5

19. In a class of 18 students, 6 are wearing jeans. What is the ratio of students wearing jeans to students *not* wearing jeans?

    A) 1:2 B) 1:3 C) 2:3 D) 2:1

20. The sum of two numbers is 12, and their product is 35. The larger of the two numbers is

    A) 8 B) 7 C) 6 D) 5

21. $(123 \times 8) + (123 \times 9) + (123 \times 10) + (123 \times 11)$ is divisible by

    A) 9 B) 8 C) 7 D) 6

22. When twice the perimeter of a square is tripled, the result is 72. What is the area of the square?

    A) 3 B) 9 C) 12 D) 16

23. Of the following numbers, which is the largest number?

    A) 1^5 B) 2^4 C) 3^3 D) 4^2

24. On every odd-numbered day in May, Dave ran for 15 minutes. On every even-numbered day in May, he ran for 44 minutes. For how many *hours* did he run in May?

    A) 15 B) 30 C) 60 D) 900

25. $5 \times \sqrt{5} \times 5 \times \sqrt{5} =$

    A) $5 \times 5 \times 25$ B) $5 \times 5 \times 5$
    C) $5 \times 5 \times 2$ D) 5×5

26. The product of two whole numbers is 30. What is the least possible value of their sum?

    A) 10 B) 11 C) 13 D) 31

27. $222 \times 66 = 333 \times 44 \times \underline{\ ?\ }$

    A) 1 B) 2 C) 3 D) 4

Go on to the next page))))▶ **6**

2009-2010 6TH GRADE CONTEST

#	Question	Answers
28.	(8 + 10 + 12) + (8 + 10 − 12) + (8 + 12 − 10) + (10 + 12 − 8) = A) (8+10+12) B) 2 × (8+10+12) C) 3 × (8+10+12) D) 4 × (8+10+12)	28. b
29.	If a whole number between 100 and 999 has three different non-zero digits, what is the least possible value of the sum of its digits? A) 7 B) 6 C) 4 D) 3	29. b
30.	In 20 years, Ed will be 31 and Di will be 35. The sum of their ages now is A) 26 B) 46 C) 86 D) 106	30. a
31.	What month is 1000 months after March? A) March B) May C) June D) July	31. d
32.	The ones digit of the product 123 × 456 × 789 is A) 1 B) 2 C) 3 D) 4	32. b
33.	An equal number of pennies, nickels, and dimes have a combined total value of $2.40. The total value of the nickels is A) 15¢ B) 50¢ C) 75¢ D) 95¢	33. c
34.	(2010 − 2005) × (2005 − 2000) × (2000 − 1995) × … × (10 − 5) × (5 − 0) = A) 5^{402} B) 5^{401} C) 5 × 402 D) 5 × 401	34. a
35.	Two equilateral triangles share sides with a square, as shown. If a side of the square has a length of 4, what is the perimeter of the figure? A) 48 B) 40 C) 32 D) 24	35. d
36.	If there are 420 students in my school, then the ratio of boys to girls in my school *cannot* be A) 3:7 B) 5:9 C) 11:14 D) 17:18	36. c
37.	300% of 300 = _?_ % of 3000 A) 10 B) 25 C) 30 D) 50	37. c
38.	Bricks weigh 3 kg or 7 kg each. Cy picks up at least one brick of each size. The total weight of bricks he picks up *cannot* be A) 21 kg B) 27 kg C) 30 kg D) 39 kg	38. a
39.	The smallest prime number that is a factor of (1 × 2 × 3 × … × 30) + 1 must be A) less than 10 B) between 10 & 20 C) between 20 & 30 D) greater than 30	39. d
40.	How many whole numbers from 1 through 500 have a 3 as the hundreds digit or ones digit, but *not* as both? A) 130 B) 140 C) 150 D) 160	40. a

The end of the contest 6

Visit our Web site at http://www.mathleague.com
Solutions on Page 129 • Answers on Page 151

SIXTH GRADE MATHEMATICS CONTEST

Math League Press, P.O. Box 17, Tenafly, New Jersey 07670-0017

2010-2011 Annual 6th Grade Contest

Tuesday, February 15 or 22, 2011

6

Instructions

- **Time** Do *not* open this booklet until you are told by your teacher to begin. You might be *unable* to finish all 35 questions in the 30 minutes allowed.

- **Scores** Please remember that *this is a contest, and not a test*—there is no "passing" or "failing" score. Few students score as high as 28 points (80% correct). Students with half that, 14 points, *should be commended!*

- **Format, Point Value, & Eligibility** Every answer is an A, B, C, or D. Write answers in the *Answers* column. A correct answer is worth 1 point. Unanswered questions get no credit. You **may** use a calculator.

Copyright © 2011 by Mathematics Leagues Inc.

2010-2011 6TH GRADE CONTEST

#	Question	Answers
1.	How many letters remain in the English alphabet after you remove all of the letters in "apple"? A) 22 B) 21 C) 20 D) 19	1. a
2.	The remainder when the sum 2011 + 201 + 20 + 2 is divided by 10 is A) 1 B) 2 C) 4 D) 8	2. c
3.	888 + 444 + 222 = 222 × _?_ A) 3 B) 6 C) 7 D) 8	3. c
4.	There are 60 parents watching a school play. If there are 3 times as many mothers as fathers watching, how many mothers are watching? A) 15 B) 20 C) 40 D) 45	4. d
5.	66 + 34 + 66 + 34 + 66 + 34 + 66 + 34 + 66 + 34 + 66 + 34 + 66 + 34 + 66 + 34 + 66 + 34 = A) 1000 B) 900 C) 660 D) 100	5. a
6.	Which of the following is *not* a factor of 123456? A) 3 B) 4 C) 6 D) 7	6. d
7.	How many different prime numbers, when multiplied by 11, have an even number as their product? A) 0 B) 1 C) 2 D) 3	7. b
8.	Which of the following must have perpendicular sides? A) rectangle B) rhombus C) trapezoid D) parallelogram	8. a
9.	$1 \times 2 + 3 \times 4 + 5 \times 6 + 7 \times 8 =$ A) 36 B) 100 C) 1256 D) 4680	9. b
10.	In 2011, the number of days in one week will be what percent of the number of days in February? A) 4% B) 7% C) 25% D) 35%	10. c
11.	Rachel has $200 in her piggy bank. She had 80% less money in her piggy bank last year. How much was in that bank last year? A) $20 B) $40 C) $80 D) $160	11. b
12.	The sum of Sid's and Sue's ages is 30. If, 8 years ago, Sid was as old as Sue was 2 years ago, then how old is Sid now? A) 12 B) 18 C) 22 D) 28	12. b
13.	$6^{2011} + 6^{2011} + 6^{2011} + 6^{2011} + 6^{2011} + 6^{2011} =$ A) 6^{2012} B) 6^{2016} C) 36^{2011} D) 36^{2016}	13. a

Go on to the next page

14.	Betaburg's bus lines are numbered starting at 33 and ending at 66, counting by 3s. How many bus lines does Betaburg have? A) 14 B) 13 C) 12 D) 11	14.
15.	Two numbers have an average of 10 and a product of 91. The smaller of the numbers is A) 7 B) 9 C) 11 D) 13	15.
16.	The sum of all whole-number factors of 32 is A) 7 B) 24 C) 30 D) 63	16.
17.	A faucet drips once every 5 minutes. In 24 hours it drips ? times. A) 36 B) 72 C) 144 D) 288	17.
18.	A rectangle is divided into four congruent regions, as shown. If the total area of the unshaded regions is 12, then the area of the original rectangle is A) 4 B) 16 C) 18 D) 48	18.
19.	What is the average of all odd numbers between 20 and 50? A) 33 B) 34 C) 35 D) 36	19.
20.	What is the difference between the greatest prime number less than 40 and the least prime number greater than 20? A) 14 B) 16 C) 18 D) 19	20.
21.	Forty minutes after 4:40 is forty minutes before A) 5:20 B) 6:00 C) 6:20 D) 6:40	21.
22.	$9^{201} \times 10^{201} = 6^{201} \times$? A) 13^{201} B) 15^{201} C) 17^{201} D) 19^{201}	22.
23.	10% of 20% of ? is 30. A) 30 B) 300 C) 600 D) 1500	23.
24.	What is the greatest common divisor of $6 \times 7 \times 8 \times 9$ and $7 \times 8 \times 9 \times 10$? A) $7 \times 8 \times 9$ B) $2 \times 7 \times 8 \times 9$ C) $6 \times 7 \times 8 \times 9$ D) $7 \times 8 \times 9 \times 10$	24.
25.	If I round 12 345 to the nearest 100, then divide by 1230, then multiply by 123, the result is A) 12340 B) 1234 C) 1230 D) 123	25.

2010-2011 6TH GRADE CONTEST

Answers

26. At least how many 2 × 2 uncut tiles must Pat use to cover a 7 × 7 section of a large floor?
    A) 13   B) 14   C) 15   D) 16

27. How many positive integers less than 1000 are multiples of 4 but do not include the digit 5?
    A) 206   B) 207   C) 208   D) 209

28. The ratio 3:5 is equal to
    A) $\frac{1}{3}:\frac{1}{5}$ B) 13:15   C) 30:50   D) (1+3):(1+5)

29. The sum of the positive factors of my number is *less* than twice my number. My number could be
    A) 12   B) 16   C) 20   D) 24

30. The figure at right is composed of squares with side-lengths of 1, 2, and 3. What is the perimeter of the figure?
    A) 18   B) 20   C) 21   D) 22

31. Brad skateboards from home to Caryn's house at 5 km/hour, then returns to his house along the same route at 7 km/hour. What is Brad's average speed, in km/hour, for the entire trip?
    A) $5\frac{5}{7}$ B) $5\frac{5}{6}$ C) 6   D) $6\frac{1}{6}$

32. (1 × 2 × 3 × ... × 50) − (1 × 2 × 3 × ... × 49) = (1 × 2 × 3 × ... × 48) × ?
    A) 49^2 B) 50^2 C) 49   D) 50

33. There are 10 players in a chess tournament. If each game is played by 2 players, and each player plays every other player exactly once, what is the total number of games played in the tournament?
    A) 100   B) 90   C) 50   D) 45

34. Pa's Pets has 26 cats and 15 fish for sale. If 6 of the fish have spots, and if the total number of spotted cats and spotted fish is 3 times the number of cats without spots, how many of these cats and fish have spots?
    A) 8   B) 9   C) 18   D) 24

35. The 2011th digit to the right of the decimal point in the decimal representation of $\frac{1}{54}$ is
    A) 0   B) 1   C) 5   D) 8

The end of the contest ☛ **6**

Visit our Web site at http://www.mathleague.com
Solutions on Page 133 • Answers on Page 152

Detailed Solutions

2006-2007 through 2010-2011

4th Grade Solutions
2006-2007 through 2010-2011

FOURTH GRADE MATHEMATICS CONTEST

Math League Press, P.O. Box 17, Tenafly, New Jersey 07670-0017

Information & Solutions

Spring, 2007

Contest Information

- **Solutions** Turn the page for detailed contest solutions (written in the question boxes) and letter answers (written in the *Answer Column* to the right of each question).

- **Scores** Please remember that *this is a contest, not a test*—and there is no "passing" or "failing" score. Few students score as high as 24 points (80% correct). Students with half that, 12 points, *deserve commendation!*

- **Answers & Rating Scale** Turn to page 138 for the letter answers to each question and the rating scale for this contest.

Copyright © 2007 by Mathematics Leagues Inc.

2006-2007 4TH GRADE CONTEST SOLUTIONS

	Answers
1. Of the choices, $3 \times 3 = 9$, choice C, has the largest value. A) $3+3 = 6$ B) $3-3 = 0$ C) $3 \times 3 = 9$ D) $3 \div 3 = 1$	1. C
2. 14 days after Monday is Monday; 2 days earlier is Saturday. A) Wednesday B) Thursday C) Friday D) Saturday	2. D
3. $55+55+55 = 44+11+44+11+55 = 44+44+(11+11+55) = 44+44+77$. A) 33 B) 44 C) 66 D) 77	3. D
4. Five dimes is worth 10 nickels. With that, I can buy 10 5¢ stamps. A) 10 B) 20 C) 25 D) 50	4. A
5. We know that The Chef has 3 dozen eggs. We know that 3 eggs are needed for each omelet. For The Chef, (# eggs) ÷ 3 = # omelets he can make. Here, (3 dozen) ÷ 3 = 1 dozen. A) 1 dozen B) 2 dozen C) 3 dozen D) 4 dozen	5. A
6. $8 \times 9 \times 10 \times 11 = (8 \times 10) \times (9 \times 11) = 80 \times 99$. A) 81 B) 88 C) 90 D) 99	6. D
7. The remainders for the listed choices are 0, 0, 1, 2, respectively. A) $463 \div 1$ B) $874 \div 2$ C) $355 \div 3$ D) $506 \div 4$	7. D
8. The word *mathematics* has 4 vowels. The word *history* has 2. A) arithmetic B) history C) reading D) computers	8. B
9. 8 pairs of bears = 16 bears have $16 \times 4 = 64$ paws = 32 pairs of paws. A) 16 B) 32 C) 48 D) 64	9. B
10. $48 \div 4 = 12 = 4 \times 3$. A) 3 B) 8 C) 12 D) 16	10. A
11. If 2 policemen use each car, the 9 cars would hold 18 policemen. But there are 19 policemen, so 1 of the cars is used by 3 policemen. A) 2 B) 3 C) 4 D) 5	11. B

Go on to the next page ▶ **4**

2006-2007 4TH GRADE CONTEST SOLUTIONS

#	Question	Answer
12.	The product is odd if and only if every factor is odd. A) 23×24 B) 24×35 C) 42×53 D) 53×45	12. D
13.	Make a first guess of 12 sandwiches each. Then, add 3 to, and subtract 3 from, the guess of 12. The results, 12+3 & 12−3, differ by 6, and 12+3 = 15. A) 9 B) 12 C) 15 D) 18	13. C
14.	Add 20 to get 120, then 120÷4 = 30. A) 120 B) 80 C) 30 D) 25	14. C
15.	Let's think in terms of how many 7s there are. Suzie has 70 7s. If she removes one of them, she'll have only 69 7s. A) 7×69 B) 7×63 C) 6×70 D) 6×63	15. A
16.	Since there are 3 sisters in Mary's family, Mary has 2 sisters. A) 2 B) 3 C) 4 D) 5	16. A
17.	To get 12×4 = 48 sides, I would need 48÷3 = 16 triangles. A) 4 B) 16 C) 36 D) 48	17. B
18.	My sunflower was 10 cm tall yesterday, so it's 20 cm tall today, 40 cm tall tomorrow, and 80 cm tall the day after tomorrow. A) 30 B) 40 C) 60 D) 80	18. D
19.	The remainder are 3, 4, 1, and 2 respectively. A) 543 = 540+3 B) 654 = 650+4 C) 876 = 875+1 D) 987 = 985+2	19. B
20.	4×8×16 = 2×(2×8)×16. A) 2 B) 4 C) 8 D) 16	20. A
21.	From 3:18 P.M. to 4:56 P.M. is 1 hr. 38 min. = 98 min. Half that is 49 min., so he finished half at 4:07 P.M. A) 3:58 P.M. B) 4:04 P.M. C) 4:07 P.M. D) 4:19 P.M.	21. C
22.	Divisibility by 6 & 20 does *not* promise divisibility by 7, 13, or 8. A) 12 = 4×3 B) 14 = 2×7 C) 26 = 2×13 D) 120 = 8×15	22. A

Go on to the next page ⏵ **4**

2006-2007 4TH GRADE CONTEST SOLUTIONS

		Answers
23.	Since $60480 = 9 \times 8 \times 7 \times 6 \times 5 \times 4$, choice C is correct. A) $680 = 17 \times \ldots$ B) $2835 = 81 \times \ldots$ C) 60480 D) $83349 = 49 \times \ldots$	23. C
24.	The sum of my 3 birds' ages is 25. In 2 years, each will be 2 years older, so the sum will then be $25+6 = 31$. A) 27 B) 29 C) 30 D) 31	24. D
25.	To find the greatest prime number less than 50, try the odd numbers less than 50. Since $49 = 7 \times 7$, try 47. It is prime. A) 49 B) 47 C) 41 D) 37	25. B
26.	Add the length and width and you get half the perimeter, 12. The length is twice the width, so the length is 8 and the width is 4. A) 2 B) 3 C) 4 D) 6	26. C
27.	My florist needs 15 flowers to make 3 special bouquets. For 6 times as many special bouquets my florist would need $6 \times 15 = 90$ flowers. A) 90 B) 54 C) 45 D) 36	27. A
28.	Every dime can be paired with a penny, so the correct answer will be divisible by 11¢. A) $1.01 B) $1.10 C) $1.11 D) $11.01	28. B
29.	There are 4 small triangles. There are also 4 triangles that are each made up of two sides of the square and one of the diagonals of the square. A) 4 B) 6 C) 8 D) 12	29. C
30.	$(100-99)+(98-97)+ \ldots +(2-1) = 1+1+ \ldots +1 = 50$. A) 25 B) 50 C) 99 D) 100	30. B

The end of the contest 4

Visit our Web site at http://www.mathleague.com

FOURTH GRADE MATHEMATICS CONTEST

Math League Press, P.O. Box 17, Tenafly, New Jersey 07670-0017

Information & Solutions

Spring, 2008

Contest Information

4

- **Solutions** Turn the page for detailed contest solutions (written in the question boxes) and letter answers (written in the *Answer Column* to the right of each question).

- **Scores** Please remember that *this is a contest, not a test*—and there is no "passing" or "failing" score. Few students score as high as 24 points (80% correct). Students with half that, 12 points, *deserve commendation!*

- **Answers & Rating Scale** Turn to page 139 for the letter answers to each question and the rating scale for this contest.

Copyright © 2008 by Mathematics Leagues Inc.

2007-2008 4TH GRADE CONTEST SOLUTIONS

#	Problem	Answer
1.	$2 \times 0 \times 0 \times 8 = 0$, since 0 times any number is 0. A) 2008 B) 16 C) 10 D) 0	1. D
2.	$4+4+4 + 4+4+4 = 12 + 12 = 2 \times 12$. A) 1×18 B) 2×12 C) 2×16 D) 4×8 4,4,4,4,4,4	2. B
3.	$444 + 555 = 999 = 666 + 333$. A) 222 B) 333 C) 444 D) 777	3. B
4.	Today is Mon., yesterday was Sun., and 2 days before Sun. was Fri. A) Friday B) Saturday C) Sunday D) Monday	4. A
5.	$3 \times 3 \times 3 \times 3 = 9 \times 9 = 81$. A) 12 B) 27 C) 81 D) 3333	5. C
6.	$4 \times 1 \times 3 \times 1 \times 2 \times 1 = 4 \times 3 \times 2 = 24$. A) 9 B) 12 C) 24 D) 39	6. C
7.	Each panda bear has 1 pair of front paws, so 40 bears have 40 pairs. A) 20 B) 40 C) 80 D) 160	7. B
8.	$91 - 19 = 72 = 81 - 9$. A) 9 B) 18 C) 19 D) 29	8. A
9.	Of the 24 bears who live in Yogi National Park, 5 were seated at a picnic table and $24 - 5 = 19$ were not. A) 5 B) 9 C) 15 D) 19	9. D
10.	$(99 \div 3) \times 3 = 33 \times 3 = 99$. A) 11 B) 33 C) 99 D) 297	10. C
11.	$(10-10) \times (10+10) = 0 \times 20 = 0$. A) 0 B) 10 C) 20 D) 100	11. A
12.	Since multiples of 100 are divisible by 4, just divide last 2 digits. A) $242 \div 4$ B) $422 \div 4$ C) $442 \div 4$ D) $452 \div 4 = 113$	12. D
13.	Since odd $\times$ odd = odd, 33×7 is an odd number. A) 66 B) 44 C) 33 D) 22	13. C

Go on to the next page ⟶ **4**

		Answers
14.	$9 \times 1 \times 9 \times 1 \times 9 = 9 \times 9 \times 9 = 729$. A) 9 B) 7 C) 3 D) 1	14. A
15.	The entire train is only 4 times as long as its engine. If the entire train is 120 m long, then the length of the engine is 120 m ÷ 4 = 30 m. A) 30 B) 80 C) 124 D) 480	15. A
16.	Since a dog has 4 legs, 600 legs ÷ (4 legs per dogs) = 150 dogs. A) 120 B) 150 C) 180 D) 300	16. B
17.	Since $51 = 3 \times 17$, the number 51 is *not* a prime. A) 11 B) 31 C) 41 D) 51	17. D
18.	To solve the equation, replace each word with the number less than 10 with which it rhymes: $(2 \times 6) + (4 \times 8) = 12 + 32 = 44$. A) 20 B) 32 C) 44 D) 64	18. C
19.	*Palindrome* years (like 1991 and 2002) read the same forwards and backwards. The years 1991 and 2002 are 11 years apart. The first palindrome year after 2002 is 2112, and 2112−2002 = 110. A) 11 B) 110 C) 111 D) 220	19. B
20.	The snowball's radius is now 10 cm long. After 1 roll its radius will be 20 cm long; after 2 rolls it will be 40 cm long; after 3 rolls it will be 80 cm long; after 4 rolls it will be 160 cm long. A) 40 B) 50 C) 80 D) 160	20. D
21.	If 9 apples weigh as much as 12 pears, and 12 pears weigh as much as 30 plums, then 9 apples weigh as much as 30 plums. A) 10 B) 19 C) 27 D) 30	21. D
22.	1 hour and 2 minutes before 3:00 P.M is 1:58 P.M. A) 1:58 P.M. B) 2:58 P.M. C) 3:58 P.M. D) 5:58 P.M.	22. A

Go on to the next page ⏵ **4**

2007-2008 4TH GRADE CONTEST SOLUTIONS

		Answers
23.	There is a net loss of 1 person after each stop. After 15 stops and starts, there are 3 people left on the bus. At the 16th stop, 3 people get off, emptying the bus. A) 18 B) 16 C) 9 D) 6	23. B
24.	Since the sum of 3 odd numbers is an odd number, the perimeter of the triangle could be 19. A) 19 B) 20 C) 22 D) 24	24. A
25.	We count $2\times 1, 2\times 2, 2\times 3, \ldots, 2\times 99$ (the 99th), and 2×100. A) 2×50 B) 2×51 C) 2×96 D) 2×99	25. D
26.	There are 4 pennies, 1 dime, and 1 nickel. A) 1 B) 2 C) 3 D) 4	26. C
27.	The 3 such whole numbers are 16, 32, and 64. A) 49 B) 45 C) 3 D) 0	27. C
28.	Put 1 couch potato on each couch. That leaves $24-18 = 6$ couch potatoes who aren't yet counted. If each sits on a different couch, there could be 6 couches with 2 occupants each. A) 9 B) 6 C) 4 D) 3	28. B
29.	The perimeter of a pentagon is the sum of its 5 side-lengths = $5\times$ length of one side = $5\times$(even number) = a multiple of 10. A) 210 B) 444 C) 666 D) 864	29. A
30.	If $1+2+\ldots+50 = 1275$, then $5\times(1+2+\ldots+50) = 5\times 1275$. A) 1525 B) 1775 C) 3825 D) 6375	30. D

The end of the contest **4**

Visit our Web site at http://www.mathleague.com

FOURTH GRADE MATHEMATICS CONTEST

Math League Press, P.O. Box 17, Tenafly, New Jersey 07670-0017

Information & Solutions

Spring, 2009

4

Contest Information

- **Solutions** Turn the page for detailed contest solutions (written in the question boxes) and letter answers (written in the *Answer Column* to the right of each question).

- **Scores** Please remember that *this is a contest, and not a test*—there is no "passing" or "failing" score. Few students score as high as 24 points (80% correct); students with half that, 12 points, *deserve commendation!*

- **Answers and Rating Scales** Turn to page 140 for the letter answers to each question and the rating scale for this contest.

Copyright © 2009 by Mathematics Leagues Inc.

2008-2009 4TH GRADE CONTEST SOLUTIONS

#	Problem	Answer
1.	The sum of 5 and 4 is 9; 6 more than 9 is 15. A) 11 B) 13 C) 15 D) 96	C
2.	Since a prime has only two different integer factors, 1 and itself, the only prime listed is 7. A) 2×3 B) 7 C) 2×4 D) 3×3	B
3.	Any number multiplied by 0 is 0. A) 0 B) 17 C) 72 D) 890	A
4.	216 is between 210 and 220 but closer to 220. A) 200 B) 210 C) 220 D) 300	C
5.	$33 + 33 + 33 = 3 + 3 + 3 + 30 + 30 + 30 = 3 + 3 + 3 + 90$. A) 3 B) 30 C) 90 D) 99	C
6.	10 minutes after 7:50 is 8:00; 20 more minutes makes it 8:20 A.M. A) 8:00 A.M. B) 8:20 A.M. C) 8:30 A.M. D) 8:50 A.M.	B
7.	$(15 - 10) + (25 - 20) + (35 - 30) + (45 - 40) = 5 + 5 + 5 + 5 = 20$. A) 5 B) 10 C) 15 D) 20	D
8.	$\$2 + 2 \times 25¢ + 2 \times 10¢ + 2 \times 5¢ = \$2 + \$0.50 + \$0.20 + \$0.10 = 280¢$. A) 280 B) 250 C) 222 D) 215	A
9.	Parentheses first: $(160) - (120) = 40$. A) 40 B) 50 C) 60 D) 70	A
10.	It takes Carlos 15 minutes each time he washes his dog. Since 1 hour = 60 minutes = 4×15 minutes, in 1 hour he washes his dog 4 times. In 3 hours, he washes his dog $3 \times 4 = 12$ times. A) 2 B) 4 C) 8 D) 12	D
11.	$18 \times 16 \times 14 = (9 \times 2) \times (8 \times 2) \times (7 \times 2) = 9 \times 8 \times 7 \times 8$. A) 2 B) 4 C) 6 D) 8	D

Go on to the next page ⟫➤ **4**

2008-2009 4TH GRADE CONTEST SOLUTIONS

	Answers
12. $2 \times 3 \times 4 \times 6 = 6 \times 24 = 12 \times 12 = 8 \times 18$. A) 6×24 B) 8×9 C) 12×12 D) 8×18	12. B
13. List multiples of 12 and 32 until a number appears on both lists. The multiples of 32 are 32, 64, 96…, and the multiples of 12 are 12, 24, 36, 48, 60, 72, 84, 96…. A) 4 B) 44 C) 96 D) 384	13. C
14. $\cancel{2} \times \cancel{3} \times \cancel{4} \times \cancel{5} \times \cancel{6} \times \cancel{7} = \cancel{6} \times \cancel{7} \times (\cancel{2} \times \cancel{4}) \times \cancel{9} \times \underline{10} \div 6$. A) 4 B) 5 C) 6 D) 7	14. C
15. 7 hours after 7:06 A.M. is 2:06 P.M.; 6 minutes later is 2:12 P.M. A) 4:12 A.M. B) 12:00 P.M. C) 2:12 P.M. D) 12:04 A.M.	15. C
16. There are 6 people ahead of Faye, making her 7th in line. The 6 behind her would be numbers 8, 9, 10, 11, 12 and 13; 13 people in line. A) 10 B) 11 C) 12 D) 13	16. D
17. Dividing 30 by 5 results in a quotient of 6. Since 30 is triple the starting number, the starting number must have been 10. A) 8 B) 10 C) 11 D) 15	17. B
18. A square, a rectangle, and a parallelogram each have 4 sides, so the sum of the number of sides is $4 + 4 + 4 = 12$. A) 11 B) 12 C) 13 D) 14	18. B
19. 16 less than 38 is 22, which is 11 more than 11. A) 11 B) 16 C) 38 D) 65	19. A
20. Twice any number divisible by 2 is divisible by 4; adding 8 creates another multiple of 4. A) 4 B) 5 C) 6 D) 7	20. A
21. 2 dozen pairs of socks = 2×12 pairs; 2×12 pairs = $2 \times 12 \times 2$ socks = 48 socks. Since 48 socks divided by 4 paws = 12, she put socks on exactly 12 cats. A) 4 B) 6 C) 12 D) 24	21. C

Go on to the next page))) ➡ **4**

2008-2009 4TH GRADE CONTEST SOLUTIONS

		Answers
22. 42 is *not* divisible by 4, so neither is 135 798 642. A) 3 B) 4 C) 6 D) 9		22. B
23. 24 = 1 × 24; 1 + 24 = 25. 24 = 3 × 8; 3 + 8 = 11. 24 = 4 × 6; 4 + 6 = 10. There is no factor pair with a product of 24 and a sum of 8. A) 25 B) 11 C) 10 D) 8		23. D
24. Since 96 = 4 × 24, there are 24 multiples of 4 between 1 and 99. The first 2 of these multiples are too small, since 4 and 8 do not have two digits. Therefore the answer is 24 − 2 = 22. A) 9 B) 18 C) 22 D) 24		24. C
25. The perimeter of a rectangle is twice the sum of its length and width; 2 × 13 = 26. A) 13 B) 26 C) 39 D) 52		25. B
26. Divide Ivan's number of stamps, 36, by 4 to get Jackie's number, 9. Divide Jackie's number, 9, by 3 to get Kenji's number, 3. A) 3 B) 4 C) 9 D) 12		26. A
27. If Lee's house is on a straight line with Mary's and Nat's and between them both, the distance between Mary's and Nat's is 11 km. A) 3 km B) 4 km C) 7 km D) 11 km		27. D
28. When 216 is divided by 53, the quotient is 4 with a remainder of 4. A) 53 B) 34 C) 20 D) 4		28. A
29. Harry is 5 years old now; he'll be 9 in 4 years. Gary is thus 3 × 9 = 27 years old. In 4 years Gary will be 27 + 4 = 31 years old. A) 10 B) 23 C) 27 D) 31		29. D
30. If 180 campers choose 2 activities each, there are 180 × 2 = 360 chosen activities. The 360 chosen activities ÷ 6 activities = 60 campers in each activity on average. A) 30 B) 60 C) 120 D) 180		30. B

The end of the contest 4

Visit our Web site at http://www.mathleague.com

FOURTH GRADE MATHEMATICS CONTEST

Math League Press, P.O. Box 17, Tenafly, New Jersey 07670-0017

Information & Solutions

Spring, 2010

Contest Information

4

- **Solutions** Turn the page for detailed contest solutions (written in the question boxes) and letter answers (written in the *Answer Column* to the right of each question).

- **Scores** Please remember that *this is a contest, and not a test*—there is no "passing" or "failing" score. Few students score as high as 24 points (80% correct); students with half that, 12 points, *deserve commendation!*

- **Answers and Rating Scales** Turn to page 141 for the letter answers to each question and the rating scale for this contest.

Copyright © 2010 by Mathematics Leagues Inc.

2009-2010 4TH GRADE CONTEST SOLUTIONS

		Answers
1.	$2 + 0 + 1 + 0 = 2 + 1 + 0.$ A) 3 B) 2 C) 1 D) 0	1. D
2.	I was 5 years old 4 years ago. I am $5 + 4 = 9$ years old now. In 3 years, I will be $9 + 3 = 12$ years old. A) 8 B) 9 C) 12 D) 13	2. C
3.	There are 3 dozen dinosaur eggs in a nest. There are $3 \times 12 = 36$ eggs. A) 15 B) 36 C) 42 D) 72	3. B
4.	The product is greater than $7 \times 120 = 840$, so choice D is the answer. A) 116 B) 130 C) 741 D) 861	4. D
5.	Since $3 \div 2$ has a remainder of 1, $3 \div 2$ is *not* a whole number. A) $3 + 2$ B) $3 - 2$ C) 3×2 D) $3 \div 2$	5. D
6.	$(10\,000 + 1) - 2 = 10\,001 - 2 = 9999.$ A) 10 002 B) 10 001 C) 9999 D) 9998	6. C
7.	Since $1 \times 2121 = 2121$, choice D is an odd number. A) 212×12 B) 12×121 C) 2×1212 D) 1×2121	7. D
8.	In 3 weeks = 21 days, it will be Tuesday. One day earlier is Monday. A) Monday B) Tuesday C) Wednesday D) Saturday	8. A
9.	Arnold the Giant Dog is holding a square board that has a perimeter of 12 m. Each side is $12 \text{ m} \div 4 = 3$ m. A) 2 m B) 3 m C) 4 m D) 5 m	9. B
10.	1 dime = 2 nickels; 10 dimes = 10×2 nickels. A) 20 B) 12 C) 5 D) 2	10. A
11.	There are 29 whole numbers from 1 to 29. Without the first 10, there are 19 numbers. A) 18 B) 19 C) 20 D) 21	11. B

Go on to the next page))))▶ **4**

2009-2010 4TH GRADE CONTEST SOLUTIONS

		Answers
12.	$40 + 80 + 120 = 40 + 200 = 240 = 4 \times 60$. A) 60 B) 50 C) 40 D) 30	12. A
13.	I have 1 dollar, 2 quarters, 3 dimes, and 4 nickels. This money is worth 100¢ + 50¢ + 30¢ + 20¢ = 200¢ = 200 pennies. A) 150 B) 175 C) 184 D) 200	13. D
14.	In (24 + 12 + 1) hours, it is 7 o'clock. A) 1 B) 4 C) 7 D) 12	14. C
15.	An elf's height increases by 10 cm each year. A gnome's height increases by 5 cm each year. Each year, the elf's height increase is 5 cm more than the gnome's. After 4 years, the elf will be 4×5 cm = 20 cm taller than the gnome. A) 10 cm B) 15 cm C) 20 cm D) 25 cm	15. C
16.	The ones digit of $(3 \times 6) + (20 \times 6) + (100 \times 6)$ is $8 + 0 + 0 = 8$. A) 2 B) 4 C) 6 D) 8	16. D
17.	Charlie turns 9 in 2010. He'll turn 21 twelve years later, in 2022. A) 2019 B) 2020 C) 2021 D) 2022	17. D
18.	All multiples of 4 are even numbers. If you add 6 to an even number, the sum remains an even number. A) even B) odd C) prime D) less than 100	18. A
19.	$(36 + 36 + 36) \div 6 = 6 + 6 + 6 = 6 \times 3$. A) 3 B) 4 C) 6 D) 9	19. A
20.	Darren dropped 6 of his flowers, but held on to 5 times as many as he dropped. He held on to $5 \times 6 = 30$ flowers, so he began with $30 + 6 = 36$. A) 30 B) 36 C) 42 D) 56	20. B
21.	There are 1000 people in a stadium. If 87 are wearing hats, then $1000 - 87 = 913$ are not wearing hats. A) 813 B) 913 C) 987 D) 1087	21. B

Go on to the next page))) ➤ **4**

2009-2010 4TH GRADE CONTEST SOLUTIONS

#	Question	Answer
22.	$(11+10)+(22+10)+(33+10)+(44+10)+(55+10) = 11+22+33+44+55+50$. A) 20 B) 30 C) 40 D) 50	22. D
23.	The sum of the digits of the number 1 100 101 is $1+1+0+0+1+0+1 = 4$. Rearranging the 1s leaves the answer unchanged. A) 4 B) 6 C) 7 D) 8	23. A
24.	$185 = 180+5$, $280 = 270+9+1$, $375 = 360+9+6$, and $470 = 450+18+2$. A) 20 R5 B) 31 R1 C) 41 R6 D) 52 R2	24. B
25.	The sum of the number of sides of a hexagon, a pentagon, and a rectangle is $6+5+4 = 15 = 3 \times 5$, the same as for 5 triangles. A) 2 B) 3 C) 4 D) 5	25. D
26.	The length of a side of each square is $8 \div 4 = 2$. Twelve such squares can fit in the rectangle. A) 5 B) 12 C) 15 D) 18	26. B
27.	Each of five brothers was born on the same date but in different years, and today is their birthday. If each is an even number of years old today but each is less than 40 years old, then their ages could not exceed 38, 36, 34, 32, and 30. The sum of those ages is 170. A) 38 B) 150 C) 170 D) 180	27. C
28.	A farm has 4 henhouses loaded with chickens. Each henhouse has 6 rows of chickens, and there are 12 chickens per row. In all, there are $4 \times (6 \times 12) = 288$ chickens. A) 22 B) 24 C) 72 D) 288	28. D
29.	If 30 is divided by 17, 18, or 19, the remainder is 13, 12, or 11. If 30 is divided by any other whole number from 1 to 16, the remainder is less than 16. A) 16 B) 12 C) 8 D) 4	29. A
30.	The difference between $(21+22+23+\ldots+38+39)$ and $(41+42+43+\ldots+59)$ is $20+20+20+\ldots+20 = 19 \times 20 = 380$. A) 190 B) 200 C) 380 D) 400	30. C

The end of the contest ☞ 4

Visit our Web site at http://www.mathleague.com

FOURTH GRADE MATHEMATICS CONTEST

Math League Press, P.O. Box 17, Tenafly, New Jersey 07670-0017

Information & Solutions

Spring, 2011

Contest Information

- **Solutions** Turn the page for detailed contest solutions (written in the question boxes) and letter answers (written in the *Answer Column* to the right of each question).

- **Scores** Please remember that *this is a contest, and not a test*—there is no "passing" or "failing" score. Few students score as high as 24 points (80% correct); students with half that, 12 points, *deserve commendation!*

- **Answers and Rating Scales** Turn to page 142 for the letter answers to each question and the rating scale for this contest.

Copyright © 2011 by Mathematics Leagues Inc.

2010-2011 4TH GRADE CONTEST SOLUTIONS

#	Problem	Answer
1.	The product of 2, 0, 1, and 1 is $2 \times 0 \times 1 \times 1 = 0$. A) 0 B) 1 C) 2 D) 4	1. A
2.	$2 + 4 + 6 + 8 = 20$. A) 8 B) 12 C) 20 D) 35	2. C
3.	Didi the dancing dog danced for exactly one week. Didi danced for 7×24 hours = 168 hours. A) 24 B) 84 C) 128 D) 168	3. D
4.	Since \$5 = 500¢, I can buy 500¢ ÷ 25¢ = 20 gumballs. A) 10 B) 20 C) 40 D) 80	4. B
5.	I have 60 marbles. If I put as many of these marbles as I can into 7 equal piles, then there are 60 ÷ 7 = 8 piles, with 4 marbles left over. A) 1 B) 4 C) 6 D) 11	5. B
6.	1000 tens minus 100 tens = 10 000 − 1000 = 9000. A) 9000 B) 9900 C) 9990 D) 9999	6. A
7.	First round 4567 to 4570. Then, 4570 − 4567 = 3. A) 3 B) 33 C) 333 D) 5433	7. A
8.	The perimeter of a rectangle is the sum of all 4 sides. Subtract 2×4 = 8 from 20. That leaves 12 for the other 2 sides. Each is 12÷2 = 6. A) 16 B) 12 C) 8 D) 6	8. D
9.	Otto did a handstand, and 1 out of every 5 pennies he had in his pockets fell out. If a total of 19 pennies fell out, Otto must have started with 5 × 19 = 95 pennies in his pockets. A) 24 B) 76 C) 95 D) 114	9. C
10.	Since 60 minutes after 5:15 is 6:15, and 15 minutes more is 6:30, the alarm should be set for 6:30. A) 6:00 B) 6:15 C) 6:30 D) 6:45	10. C
11.	The product of odd numbers is always odd. A) even B) odd C) prime D) greater than 20	11. B

Go on to the next page ▶ **4**

2010-2011 4TH GRADE CONTEST SOLUTIONS   Answers

12. Any number divisible by 12 = 2×2×3 and 42 = 2×3×7 is divisible by 2×2×3×7 = 84. Since 84 is the least whole number that is divisible by 12 and 42, there are 84 ants. A) 54 B) 84 C) 124 D) 142	12. B
13. 15+15+15+15+15+15 = 15×6 = 90 = 5×18. A) 18 B) 15 C) 6 D) 5	13. A
14. For every 4 animals in this zoo, 3 are females. There are a total of 24 animals in the zoo, so there are 6 groups of 4 animals. Each of these 6 groups has 3 females, so there are 3×6 = 18 females. A) 21 B) 18 C) 16 D) 6	14. B
15. If the radius of my circle is 4, its diameter is 2×4 = 8. The length of each side of my triangle is 2×8 = 16. Its perimeter is 3×16 = 48. A) 6 B) 24 C) 32 D) 48	15. D
16. $100 + $20 + $3 + $4.00 + $0.50 + $0.06 = $127.56. A) $123.46 B) $124.56 C) $126.96 D) $127.56	16. D
17. From 1 to 999 is 999 numbers. Without the first 9 there are 990 in all. A) 986 B) 988 C) 990 D) 992	17. C
18. Maddie earned $8 per hour helping her friend move. She worked 5 days for 8 hours each day. Maddie earned $8×5×8 = $320. A) $40 B) $64 C) $320 D) $448	18. C
19. 4 pennies + 1 dime = 14¢; 2 pennies + 3 dimes = 32¢; 1 penny + 2 nickels + 2 quarters = 61¢. A) 14¢ B) 32¢ C) 43¢ D) 61¢	19. C
20. Of the choices, 90 is divisible by the most different numbers: 1, 2, 3, 5, 6, 9, 10, 15, 18, 30, 45 and 90. A) 112 B) 90 C) 30 D) 8	20. B
21. $1 = 4 quarters and $5 = 100 nickels; that's 104 coins. A) 8 B) 29 C) 104 D) 125	21. C

Go on to the next page ⟩⟩⟩➤ 4

2010-2011 4TH GRADE CONTEST SOLUTIONS

		Answers
22.	If the same number fell on plates as on his face, each would have 40. But 16 more fell on plates than on his face, so add half of 16 to 40 to get 48 pancakes on plates. A) 24 B) 32 C) 48 D) 56	22. C
23.	A number divisible by 12 & 5 is divisible by a product of their factors, such as 3×5. A) 21 B) 15 C) 9 D) 8	23. B
24.	123÷4 = 30 with remainder 3. The sum of 30 and 3 is 33. A) 18 B) 24 C) 30 D) 33	24. D
25.	Choice A is not the sum of three different even whole numbers. A) 4 B) 12 = 2+4+6 C) 16 = 2+4+10 D) 20 = 4+6+10	25. A
26.	The two largest divisors of 28 are 28 and 14. Their difference is 14. A) 27 B) 14 C) 7 D) 6	26. B
27.	There are two prime numbers between 10 and 15: 11 and 13. A) 10 and 15 B) 20 and 25 C) 26 and 30 D) 45 and 50	27. A
28.	The first four stripes on a wall with 100 stripes are red, blue, white and purple, in that order. The 4th, 8th, 12th, ... , 52nd, and 56th stripes are all purple. The 55th stripe (one before the 56th) is white. A) red B) blue C) white D) purple	28. C
29.	In 20 years, Li will be 3 times his age now, so 20 must be twice his age now. Thus, Li is 10 now and will be 20 in 10 years. A) 10 B) 20 C) 30 D) 50	29. B
30.	Harry's beard was 4 cm long 4 days ago. It was 8 cm 3 days ago, 16 cm 2 days ago, and 32 cm 1 day ago. Today it is 64 cm. Tomorrow it will be 128 cm. A) 16 cm B) 32 cm C) 64 cm D) 128 cm	30. D

The end of the contest 4

Visit our Web site at http://www.mathleague.com

5th Grade Solutions

2006-2007 through 2010-2011

FIFTH GRADE MATHEMATICS CONTEST

Math League Press, P.O. Box 17, Tenafly, New Jersey 07670-0017

Information & Solutions

Spring, 2007

Contest Information

5

- **Solutions** Turn the page for detailed contest solutions (written in the question boxes) and letter answers (written in the *Answer Column* to the right of each question).

- **Scores** Please remember that *this is a contest, not a test*—and there is no "passing" or "failing" score. Few students score as high as 24 points (80% correct). Students with half that, 12 points, *deserve commendation!*

- **Answers & Rating Scale** Turn to page 143 for the letter answers to each question and the rating scale for this contest.

Copyright © 2007 by Mathematics Leagues Inc.

2006-2007 5TH GRADE CONTEST SOLUTIONS

		Answers
1.	$(15-14) + (13-12) + (11-10) = 1 + 1 + 1 = 3$ A) 0 B) 1 C) 3 D) 6	1. C
2.	My class has 20 boys and 40 girls. That's a total of 60 students. A) 10 students B) 30 students C) 40 students D) 60 students	2. D
3.	Since 8 is divisible by 4, choice C is correct. A) 5 B) 7 C) 8 D) 10	3. C
4.	One-third of $48 = 48 \div 3 = 16 = 2 \times 8$. A) 2 B) 4 C) 8 D) 16	4. C
5.	If it costs 75¢ to airmail a postcard from the United States to Australia, then it costs 75¢ × 50 = $37.50 to mail 50 such postcards. A) $37.50 B) $50 C) $75 D) $3750	5. A
6.	24×60 minutes = 24×1 hour = 24 hours = 1 day. A) 1 day B) 60 hours C) 24 days D) 60 days	6. A
7.	A quarter more than a nickel is 30¢, and a nickel less is 25¢. A) penny B) nickel C) dime D) quarter	7. D
8.	If 23 is divided by 3, the quotient is 7 and the remainder is 2. A) 13 B) 23 C) 33 D) 43	8. B
9.	Each Scout toasted 1 marshmallow for each of the other 2 Scouts in the group. So each scout toasted 2 marshmallows. Since there were 30 scouts, altogether they toasted $2 \times 30 = 60$ marshmallows. A) 30 B) 60 C) 90 D) 120	9. B
10.	$10\,100 \div 100 = 101$ and $10\,100 \div 10 = 1010$. A) 1 010 010 B) 101 010 C) 100 010 D) 10 100	10. D

Go on to the next page ⟶ **5**

2006-2007 5TH GRADE CONTEST SOLUTIONS

#	Question	Answer
11.	The 3 factors of 27 that are factors of 72 are 1, 3, and 9. A) 9 B) 4 C) 3 D) 2	11. C
12.	Divide choice D by 7 in your head: $728714 \div 7 = 104102$. A) 777 352 B) 770 143 C) 742 001 D) 728 714	12. D
13.	Of every 4 cats, 3 are males and 1 is female. Triple this for 12 cats. A) 9 B) 8 C) 4 D) 3	13. A
14.	Since $216 \div 72 = 3$, I can process 216 photos in 3×2 hrs. = 6 hrs. A) 3 hours B) 6 hours C) 9 hours D) 12 hours	14. B
15.	$27 \times 27 \times 27 = 9 \times 3 \times 3 \times 9 \times 27$. A) 3 B) 9 C) 27 D) 81	15. C
16.	If 5 of us split 60 coins equally, we each got $60 \div 5 = 12$ coins. A) 20 B) 15 C) 12 D) 10	16. C
17.	Class is 2 mins. less than 2 hrs.; it's half over in 1 min. less than 1 hr. A) 2:22 PM B) 2:24 PM C) 2:29 PM D) 2:31 PM	17. A
18.	I need (6×3) pages + (8×2) pages = $(18+16)$ pages = 34 pages. A) 6 B) 14 C) 34 D) 48	18. C
19.	$(11 \times 12 \times 13 \times 14) \div (1 \times 2 \times 12) = (11 \times 13 \times 14) \div 2 = 11 \times 13 \times 7$. A) $6 \times 11 \times 13$ B) $7 \times 11 \times 13$ C) $11 \times 13 \times 14$ D) $10 \times 10 \times 10 \times 10$	19. B
20.	My pet turtle walks 20×60 cm = 1200 cm in 20 days. A) 12 cm B) 30 cm C) 120 cm D) 1200 cm	20. D
21.	Since $(48 \text{ cm}) \div (4 \text{ cm}) = 12$, we need 12 pencils. A) 3 B) 4 C) 8 D) 12	21. D

Things moving too fast for you?

Go on to the next page ⏵ **5**

2006-2007 5TH GRADE CONTEST SOLUTIONS

		Answers
22.	A radius of the large circle = 3 radii of the small circle, so a radius of the large circle is 12 cm. A) 6 cm B) 8 cm C) 12 cm D) 24 cm	22. C
23.	Add 5 to 27 to get 32, twice my age now. I must be 16 now, so 3 years ago I was 13. A) 13 B) 15 C) 19 D) 24	23. A
24.	Check the odd digits only: 1, 3, and 9 are factors of 123 456 789. A) 0 B) 3 C) 5 D) 7	24. B
25.	Since 600 = 24 × 25, choice D is correct. A) 597 B) 598 C) 599 D) 600	25. D
26.	In 6 minutes, Ann hiccups 20 times and Al sneezes 15 times. Therefore, Ann hiccups 5 times more than Al sneezes. In 60 minutes, that is 50 fewer sneezes than hiccups. A) 5 B) 10 C) 50 D) 100	26. C
27.	If the sum of 2 whole numbers is 12, and their product is 32, then the numbers are 8 and 4; 8−4 = 4. A) 6 B) 4 C) 2 D) 0	27. B
28.	Since 3×6×9 = 2×81, it's divisible by 81 = 9×9. A) 3 B) 9 C) 3×9 D) 9×9	28. D
29.	Each of the star's "points" is a triangle. Each long line segment lies on a different triangle. Adding: 5+5 = 10. A) 10 B) 9 C) 8 D) 7	29. A
30.	(100−50) + (99−49) + ... + (51−1) = 50×50 = 2500. A) 250 B) 2500 C) 2525 D) 2550	30. B

The end of the contest ✎ **5**

Visit our Web site at http://www.mathleague.com

FIFTH GRADE MATHEMATICS CONTEST

Math League Press, P.O. Box 17, Tenafly, New Jersey 07670-0017

Information & Solutions

Spring, 2008
Contest Information

5

- **Solutions** Turn the page for detailed contest solutions (written in the question boxes) and letter answers (written in the *Answer Column* to the right of each question).

- **Scores** Please remember that *this is a contest, not a test*—and there is no "passing" or "failing" score. Few students score as high as 24 points (80% correct). Students with half that, 12 points, *deserve commendation!*

- **Answers & Rating Scale** Turn to page 144 for the letter answers to each question and the rating scale for this contest.

Copyright © 2008 by Mathematics Leagues Inc.

2007-2008 5TH GRADE CONTEST SOLUTIONS   Answers

1. $(110-10) + (120-20) + (130-30) = 100 + 100 + 100 = 300.$
   A) 100   B) 200   C) 300   D) 400

   1. C

2. Notice that 30 is 12 more than 18. With 12 more tickets, I can take $12 \div 3 = 4$ more rides.
   A) 4   B) 6   C) 10   D) 12

   2. A

3. The product of any number and 0 is 0.
   A) 0   B) 42   C) 220   D) 12 100

   3. A

4. Since $48 \div 3 = 16$, I have 16 Bigfish.
   A) 16   B) 24
   C) 32   D) 36

   4. A

5. 10° colder than 45° = 45° − 10° = 35°.
   A) 34°   B) 35°   C) 54°   D) 55°

   5. B

6. 1 week + 20 days = 27 days = (21+6) days = 3 weeks + 6 days.
   A) 0   B) 6   C) 7   D) 13

   6. B

7. My play lasts $3 \times (4 \times 10 \text{ minutes}) = 120$ minutes = 2 hours.
   A) 43 minutes   B) 1 hr. 10 mins.   C) 1 hr. 20 mins.   D) 2 hours

   7. D

8. An octagon has 8 sides. That's twice as many sides as a square's 4 sides.
   A) triangle   B) square
   C) pentagon   D) hexagon

   8. B

9. (Odd #) × (odd #) = odd #.
   A) 23 × 24   B) 23 × 25
   C) 24 × 25   D) 24 × 26

   9. B

10. $989 − 98 = 989 − 100 + 2 = 889 + 2 = 891.$
    A) 881   B) 887   C) 891   D) 898

    10. C

11. Increasing 4 numbers by 3 each increases their sum by $4 \times 3 = 12$.
    A) 3   B) 4   C) 9   D) 12

    11. D

Go on to the next page ⟶ 5

2007-2008 5TH GRADE CONTEST SOLUTIONS   Answers

12. Add the digits. If the sum isn't a multiple of 3, neither is the #. A) 9+3+6 = 18 B) 9+5+7 = 21 C) 6+4+5 = 15 D) 6+2+9 = 17	12. D
13. Divide $72 into 4 equal shares: $72÷4 = $18. I have 3×$18 = $54. A) $54 B) $48 C) $24 D) $18	13. A
14. The E & W steps cancel each other, so use N & S steps: my dog took 4 steps south and 6 steps north, so he was 6−4 steps = 2 steps north of his starting point. A) 2 B) 3 C) 4 D) 5	14. A
15. Average = $\frac{2\times 2 + 6\times 6}{2+6} = \frac{4+36}{8} = 5$. A) 1 B) 4 fours C) 5 D) 8	15. C
16. Since 2007/7 = 286R5, the next multiple is 7×287 = 2009. A) 2008 B) 2009 C) 2014 D) 2016	16. B
17. Of 60 flamingos, 40 stood on 2 legs and 20 stood on 1. Altogether, these flamingos stood on [(40×2)+20] legs = 100 legs. A) 80 B) 90 C) 100 D) 120	17. C
18. Use only the ones' digit! Divide each by 5, as shown below. A) 8÷5 = 1r3 B) 7÷5 = 1r2 C) 6÷5 = 1r1 D) 4÷5 = 0r4	18. D
19. $8 = 80 dimes, and 80 quarters is worth $$\frac{80}{4}$ = $20. A) $8 B) $20 C) $25 D) $40	19. B
20. The sum of the odd numbers is 7+9+11 = 27. The sum of the even numbers is 6+8+10 = 24. Finally, 27 is 3 more than 24. A) 3 B) 4 C) 5 D) 6	20. A
21. Six years ago, the age sum was 3×6 = 18 less; it was 48−18 = 30. A) 24 B) 30 C) 36 D) 42	21. B
22. The value at the end of 1 day is $64; at the end of 2 days it's $32; at the end of 3 days it's $16; and at the end of 4 days, it's $8. A) $32 B) $16 C) $8 D) $4	22. C

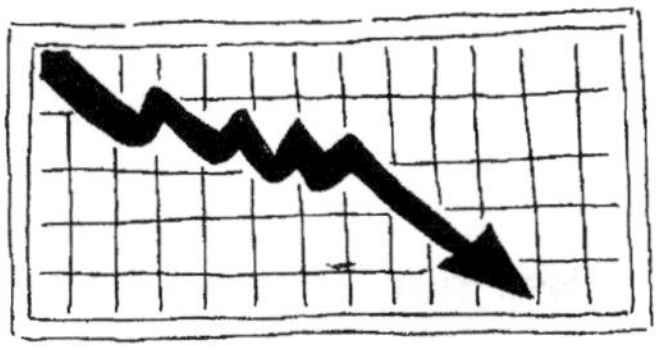

Go on to the next page ⟶ 5

2007-2008 5TH GRADE CONTEST SOLUTIONS

		Answers
23.	The greatest common factor of 36 & 63 and 45 & 54 is 9. A) 15 & 51 B) 24 & 42 C) 35 & 53 D) 45 & 54	23. D
24.	The side-length of the large square is $36 \div 4 = 9$. For the small square, it's $9 \div 3 = 3$. Perim. $= (4 \times 9)+(4 \times 3) = 48$. A) 40 B) 45 C) 48 D) 96	24. C
25.	If the difference between two basketball scores is even, then their sum must also be even. A) 0 B) 1 C) 2 D) 4	25. B
26.	5 AM is 12 hr = 720 min before 5 PM; 500 min after 5 AM is $720 - 500 = 220$ min before 5 PM. A) 240 B) 220 C) 180 D) 100	26. B
27.	The sum of the 48 numbers is 48×64. That's like 48 64s. Now remove one. What remains is 47×64. A) 47×64 B) 48×63 C) 47×63 D) 63×64	27. A
28.	The only numbers called twice are multiples of both 4 and 6. All such numbers are multiples of 12. There are $600 \div 12 = 50$ such numbers. A) 1 B) 12 C) 25 D) 50	28. D
29.	Work backwards: 24 guzzles $= 2 \times 12$ guzzles $= 5 \times 12$ gulps $= 60$ gulps $= 3 \times 20$ gulps $= 8 \times 20$ sips $= 160$ sips. A) 24 B) 72 C) 160 D) 240	29. C
30.	The 10 triangles are (1), (1+2), (1+2+3), (1+2+3+4), (2), (2+3), (2+3+4), (3), (3+4), and (4). A) 5 B) 6 C) 8 D) 10	30. D

The end of the contest ✍ **5**

Visit our Web site at http://www.mathleague.com

FIFTH GRADE MATHEMATICS CONTEST

Math League Press, P.O. Box 17, Tenafly, New Jersey 07670-0017

Information & Solutions

Spring, 2009

Contest Information

5

- **Solutions** Turn the page for detailed contest solutions (written in the question boxes) and letter answers (written in the *Answer Column* to the right of each question).

- **Scores** Please remember that *this is a contest, and not a test*—there is no "passing" or "failing" score. Few students score as high as 24 points (80% correct); students with half that, 12 points, *deserve commendation!*

- **Answers and Rating Scales** Turn to page 145 for the letter answers to each question and the rating scale for this contest.

Copyright © 2009 by Mathematics Leagues Inc.

2008-2009 5TH GRADE CONTEST SOLUTIONS

		Answers
1.	$(2+2)+(7+2)+(12+2) = 2+7+12+6$. A) 2 B) 3 C) 6 D) 7	1. C
2.	Ivan's Ice Cream Stand sold only 6 cherry, 5 vanilla, 3 lime, and 2 chocolate cones. The total number of "other" cones sold is $5+3+2 = 10$; $10-6 = 4$ fewer cherry cones were sold. A) 1 B) 2 C) 3 D) 4	2. D
3.	Each angle of a square is a right angle, which measures 90 degrees. A) 45 B) 60 C) 90 D) 100	3. C
4.	In each case except A the product is 18 000; the product in A is 1800. A) 60×30 B) 6×3000 C) 300×60 D) 3×6000	4. A
5.	$(9 \times 6 \times 3) \div (3 \times 2 \times 1) = (6 \times 9 \times 3) \div (6 \times 1) = (9 \times 3) \div 1 = 27$. A) 3 B) 6 C) 9 D) 27	5. D
6.	Only the ones' digits matter: 4×6 gives a 4; then 4×8 gives a 2. A) 2 B) 4 C) 6 D) 8	6. A
7.	3 dozen = $3 \times 12 = 36 = 18 \times 2 = 18$ pairs. A) 36 B) 18 C) 12 D) 6	7. B
8.	Barry the bullfrog weighs four times as much as Taylor the tree frog. If Barry weighs 200 g, then Taylor weighs 200 g $\div 4 = 50$ g. A) 40 B) 50 C) 250 D) 800	8. B
9.	A quadrilateral is a four-sided figure; a pentagon has 5 sides. A) pentagon B) rectangle C) square D) parallelogram	9. A
10.	13 dimes + 13 pennies = $(13 \times 10¢) + (13 \times 1¢) = 130¢ + 13¢ = 143¢$. A) 130 B) 133 C) 143 D) 146	10. C
11.	Trial and error will show that the two whole numbers with a sum of 16 and a difference of 4 are 10 and 6; their product is $10 \times 6 = 60$. A) 56 B) 60 C) 64 D) 78	11. B
12.	$(24+30)$ hrs. = 54 hrs.; 3×24 hrs. = 72 hrs.; 72 hrs. $-$ 54 hrs. = 18 hrs. A) 6 B) 12 C) 18 D) 24	12. C

Go on to the next page ⟩⟩⟩▶ **5**

2008-2009 5TH GRADE CONTEST SOLUTIONS

#	Problem	Answer
13.	Since 56 × 4 = 224, which is less than 240, 240 ÷ 56 is greater than 4. A) 240 ÷ 56 B) 168 ÷ 42 C) 224 ÷ 58 D) 138 ÷ 38	A
14.	2 hours = 2 × 60 min. = 120 min.; 120 min. ÷ 10 min. = 12; he bikes 12 × 3 km = 36 km. A) 9 km B) 18 km C) 24 km D) 36 km	D
15.	The thousands' digit is the fourth from the right; in 12 345 it is 2. Twice 2 is 4. A) 2 B) 4 C) 6 D) 8	B
16.	(4 × 4) × (4 × 4) × (4 × 4) = 16 × 16 × 16. A) 16 × 16 × 16 B) 16 + 16 + 16 C) 16 × 16 D) 16 + 16	A
17.	A prime number has exactly two different positive factors. A) 1 × 42, 2 × 21 B) 1 × 43 C) 1 × 49, 7 × 7 D) 1 × 51, 3 × 17	B
18.	An equilateral triangle has 3 equal sides, so if its perimeter is 24, each side has length 24 ÷ 3 = 8. A) 4 B) 6 C) 8 D) 12	C
19.	If today is the 10th, tomorrow is the 11th; 3 days later is the 14th. The day before yesterday was the 8th. The difference is 6 days. A) 4 B) 5 C) 6 D) 7	C
20.	4000 × 3000 = 12 000 000; the first two digits are not zeros. A) 1 B) 2 C) 6 D) 8	B
21.	If there are a total of 110 people in line, subtract those behind Frank and Frank himself: 110 − 76 − 1 = 33, the number in front of Frank. A) 33 B) 34 C) 35 D) 36	A
22.	Purchasing 7 pens includes the first pen and 6 others. Those pens would cost 80¢ + (65¢ × 6) = 80¢ + 390¢ = 470¢. Gina's purchase would cost $4.70. A) $4.25 B) $4.40 C) $4.55 D) $4.70	D
23.	Since 9 glorks = 5 blorks, 10 × 9 glorks = 10 × 5 blorks = 50 blorks. A) 10 B) 18 C) 45 D) 50	D

Go on to the next page ⟶ 5

2008-2009 5TH GRADE CONTEST SOLUTIONS

	Answers

24. Since the original rectangle has a length of 6 and a width of 3, and since the lengths of the four sides of each square are equal, each square must have sides of length 3. The perimeter of each is 4 × 3 = 12.

A) 3   B) 6   C) 9   D) 12

24. D

25. 10 × (2 × 15) × (2 × 25) = (15 × 25) × (10 × 2 × 2) = 15 × 25 × 40.

A) 35   B) 40   C) 45   D) 50

25. B

26. Choose any number greater than 600 000 for the first day's attendance. Suppose 700 000 people attended on the first day. That would mean 500 000 people attended on the third day. The average attendance for the three days would be (700 000 + 600 000 + 500 000) ÷ 3 = 600 000. Choose any other number and you get the same answer.

A) 200 000   B) 300 000   C) 600 000   D) 1 200 000

26. C

27. If the hundreds' digit is 1, there are 2 numbers (101 and 110); if it is 2, there are 3 (202, 211, 220), etc.; 2 + 3 + 4 + 5 + 6 + 7 + 8 + 9 + 10 = 54.

A) 18   B) 20   C) 36   D) 54

27. D

28. She must have 3 pennies and 7 other coins. If she had only 1 quarter, she'd need 5 dimes and 1 nickel, but the number of quarters would be the same as the number of nickels. She must have 2 quarters, and thus 1 dime and 4 nickels. The number of nickels is greatest.

A) pennies   B) nickels   C) dimes   D) quarters

28. B

29. At this rate, Jackie would be the first student to arrive 2 × 5 out of 2 × 8 days, 3 × 5 out of 3 × 8 days, ... , 5 × 5 out of 5 × 8 days. In 40 days, she would be first to arrive 25 times.

A) 8   B) 13   C) 20   D) 25

29. D

30. There are 5 ways to arrange the 100 square tiles into a rectangle: 1 by 100, 2 by 50, 4 by 25, 5 by 20, and 10 by 10. The perimeters of those arrangements are 202 m, 104 m, 58 m, 50 m, and 40 m.

A) 40 m   B) 100 m   C) 202 m   D) 404 m

30. A

The end of the contest 5

Visit our Web site at http://www.mathleague.com

FIFTH GRADE MATHEMATICS CONTEST

Math League Press, P.O. Box 17, Tenafly, New Jersey 07670-0017

Information & Solutions

Spring, 2010

Contest Information

5

- **Solutions** Turn the page for detailed contest solutions (written in the question boxes) and letter answers (written in the *Answer Column* to the right of each question).

- **Scores** Please remember that *this is a contest, and not a test*—there is no "passing" or "failing" score. Few students score as high as 24 points (80% correct); students with half that, 12 points, *deserve commendation!*

- **Answers and Rating Scales** Turn to page 146 for the letter answers to each question and the rating scale for this contest.

Copyright © 2010 by Mathematics Leagues Inc.

2009-2010 5TH GRADE CONTEST SOLUTIONS

#	Question	Answer
1.	There are 9 tuba players ahead of Ira in line, and 20 behind him. Counting Ira, there are 9 + 20 + 1 = 30 tuba players in the line. A) 28 B) 29 C) 30 D) 31	1. C
2.	$(2 \times 12 \times 5) \div 12 = 2 \times 5 = 10$. A) 6 B) 8 C) 10 D) 12	2. C
3.	Since 30 m ÷ 4 m has a quotient of 7 and a remainder of 2, Rachel can cut at most 7 pieces that are 4 m long. A) 6 B) 7 C) 8 D) 9	3. B
4.	Sam finds 4 shells each minute at the beach. Since there are 60 minutes in an hour, she finds 4 × 60 = 240 shells in an hour. A) 15 B) 20 C) 64 D) 240	4. D
5.	$(32 \div 4) + (48 \div 6) + (64 \div 8) = 8 + 8 + 8 = 8 \times 3$. A) 3 B) 6 C) 8 D) 12	5. A
6.	$9 \times 9 = 81 = 90 - 9$. A) 999 ÷ 9 B) 90 + 9 C) 90 − 9 D) 99	6. C
7.	Tom, Dick, and Harry go to the movies together. Tom pays for all 3 tickets with a $50 bill. If he gets $18.50 in change, the 3 tickets cost $50.00 − $18.50 = $31.50. The cost of one ticket is $31.50 ÷ 3 = $10.50. A) $10.50 B) $11.50 C) $12.50 D) $18.50	7. A
8.	1 + 0 + 5 + 0 is divisible by 3, so 1050 must also be divisible by 3. A) 1050 B) 2024 C) 3058 D) 4022	8. A
9.	When 4 of Tom's friends paint his fence, it takes 12 hours to paint it. With twice as many workers, it would take half as long, 6 hours. A) 4 B) 6 C) 8 D) 24	9. B
10.	It's 10 mins. to 3 P.M. and 15 more to 3:15. A) 25 B) 45 C) 65 D) 75	10. A
11.	The product of 2004 and 2005 has a ones digit of 0. Therefore, the entire product also has a ones digit of 0. A) 8 B) 6 C) 4 D) 0	11. D

Go on to the next page))) ➔ 5

2009-2010 5TH GRADE CONTEST SOLUTIONS

#	Question	Answer
12.	The value is (25 × 25¢)+(10 × 10¢)+(5 × 5¢)+(1 × 1¢) = (625+100+25+1)¢. A) $7.01 B) $7.16 C) $7.51 D) $8.11	12. C
13.	Since 300 ÷ 8 has a quotient of 37, the desired multiple is 8 × 37 = 296. A) 298 B) 296 C) 294 D) 292	13. B
14.	2010 ÷ 30 = 67, 67 − 3 = 64, and 64 = 4 × 16. A) 7 × 10 B) 6 × 12 C) 5 × 14 D) 4 × 16	14. D
15.	If 4 out of 5 dentists recommend sugarless gum, then 1 out of 5 do not recommend it. Since 180 ÷ 5 = 36, there are 36 groups of 5 dentists each, and 1 dentist from each group does not recommend it. A) 9 B) 36 C) 72 D) 144	15. B
16.	20 × 100 + 20 × 10 + 20 = 2000 + 200 + 20 = 2220. A) 222 B) 2022 C) 2040 D) 2220	16. D
17.	161 616 169 ÷ 8 = 20 202 021 with remainder 1. A) 1 B) 3 C) 5 D) 7	17. A
18.	Since the same number of students took the contest at each school, the average score is the average of 10 and 6. The average of 10 and 6 is (10 + 6) ÷ 2 = 16 ÷ 2 = 8. A) 10 B) 8 C) 7 D) 6	18. B
19.	7 hours before 7 P.M. is noon, and 7 hours before noon is 5 A.M. A) 5 A.M. B) 7 A.M. C) 2 P.M. D) 7 P.M.	19. A
20.	2×2 blurps = 2×5 blaps, so 1 blorp = 10 blaps and 6×1 blorps = 6×10 blaps. A) 16 B) 24 C) 48 D) 60	20. D
21.	Gavin is a monkey and Hal is his owner. Hal's age is 7 years greater than 3 times Gavin's age. If Gavin is 11, then Hal is 3 × 11 + 7 = 40. The difference between their ages is 40 − 11 = 29. A) 15 B) 18 C) 29 D) 51	21. C
22.	The numbers are each about half of 4019; they are 2009 and 2010. The difference between any pair of consecutive whole numbers is always 1. A) 1 B) 9 C) 2009 D) 2010	22. A

Go on to the next page 5

2009-2010 5TH GRADE CONTEST SOLUTIONS

#	Question	Answer
23.	A hexagon has 6 sides. The others all have 4 sides. A) hexagon B) rhombus C) trapezoid D) parallelogram	23. A
24.	Together, Alice and Bob write 20 letters. Together, Charlie, Dan, and Evelyn write 60 letters. The average number of letters these five people write is 80 ÷ 5 = 16. A) 18 B) 16 C) 15 D) 12	24. B
25.	From 0 to 6, 2, 3, and 5 are prime (but 1 is not a prime). A) 0 and 6 B) 6 and 12 C) 12 and 18 D) 18 and 24	25. A
26.	Jesse and James each had $10 to spend on school supplies. Jesse spent his $10 on pens that cost $0.50 each. James spent his $10 on pencils that cost $0.20 each. Jesse bought $10 ÷ $0.50 = 20 pens and James bought $10 ÷ $0.20 = 50 pencils. The difference is 50 − 20 = 30. A) 10 B) 25 C) 30 D) 50	26. C
27.	Each side of a small square is 8 ÷ 4 = 2. The area of rectangle R is 6 × 4 = 24. A) 10 B) 12 C) 20 D) 24	27. D
28.	When you divide 123 by 4, the quotient is 30 and the remainder is 3. Now divide 30 by 3 to get a quotient of 10. A) 5 B) 10 C) 15 D) 23	28. B
29.	The numbers beginning with 1 are 1234, 1243, 1324, 1342, 1423, and 1432. There are also 6 numbers beginning with 2 or 3 or 4. A) 6 B) 12 C) 18 D) 24	29. D
30.	Any multiple of 100 is divisible by 4. If the 2-digit number formed by the last 2 digits of each original number is divisible by 4, then so is the original number. Since 92 is divisible by 4, 67■92 is also divisible by 4. A) 24■58 B) 53■49 C) 25■74 D) 67■92	30. D

The end of the contest 5

Visit our Web site at http://www.mathleague.com

FIFTH GRADE MATHEMATICS CONTEST

Math League Press, P.O. Box 17, Tenafly, New Jersey 07670-0017

Information & Solutions

Spring, 2011

Contest Information

5

- **Solutions** Turn the page for detailed contest solutions (written in the question boxes) and letter answers (written in the *Answer Column* to the right of each question).

- **Scores** Please remember that *this is a contest, and not a test*—there is no "passing" or "failing" score. Few students score as high as 24 points (80% correct); students with half that, 12 points, *deserve commendation!*

- **Answers and Rating Scales** Turn to page 147 for the letter answers to each question and the rating scale for this contest.

Copyright © 2011 by Mathematics Leagues Inc.

2010-2011 5TH GRADE CONTEST SOLUTIONS

		Answers
1.	2010 + 2020 + 2030 = 6000 + (10 + 20 + 30). A) 30 B) 50 C) 60 D) 80	1. C
2.	I have 50 fish. If I want to put 2 fish in each bowl, I need 50÷2 = 25 fish bowls. A) 25 B) 52 C) 100 D) 502	2. A
3.	10 + 8 × 6 − 4 ÷ 2 = 10 + 48 − 2 = 56. A) 18 B) 27 C) 52 D) 56	3. D
4.	4 quarters + 4 dimes + 4 nickels + 4 pennies = $1+40¢+20¢+4¢ = $1.64. A) $1.64 B) $2.44 C) $3.64 D) $4.44	4. A
5.	Since 300 × 3 is even and divisible by 3, it is a multiple of 6. A) 500 × 2 B) 400 × 4 C) 300 × 3 D) 200 × 8	5. C
6.	A pentagon has 5 sides, and 5 is an odd number. A) hexagon B) pentagon C) rectangle D) rhombus	6. B
7.	One of my two brothers is 4 years older than the other. If they were the same age, they'd each be 19. Thus, one is 17 and the other is 21. A) 17 B) 21 C) 23 D) 27	7. B
8.	If Lenny has 63 books, and he has 2 hardcover books for every 7 softcover books, 2 of every 9 books are hardcover books. Since there are 7 groups of 9 books in all, there are 7 × 2 = 14 hardcover books. A) 14 B) 18 C) 43 D) 49	8. A
9.	The desired number must be even, so it is either 2224 or 3350. Only 3350 has a remainder of 2 when divided by 4. A) 2224 B) 3350 C) 4481 D) 5523	9. B
10.	Rose is 150 cm tall. Quentin is 10 cm taller than Rose, so Quentin is 160 cm tall. Sam is 4 cm shorter than Rose, so Sam is 146 cm tall. Their average height is (150+160+146)÷3 = 152 cm. A) 148 cm B) 150 cm C) 152 cm D) 154 cm	10. C
11.	The only two whole numbers whose product is 36 and whose sum is 13 are 4 and 9. Their difference is 9 − 4 = 5. A) 8 B) 7 C) 6 D) 5	11. D

Go on to the next page ⟩⟩⟩➧ **5**

2010-2011 5TH GRADE CONTEST SOLUTIONS

#	Solution	Answer
12.	7 hours before 7 P.M. is noon. 60 minutes before noon is 11 A.M. 17 minutes before 11A.M. is 10:43 A.M. That means they must arrive at 10:43 A.M. A) 10:43 A.M. B) 11:17 A.M. C) 12:07 P.M. D) 12:17 P.M.	12. A
13.	60 more than 0 is 60; the product is 6 × 0. A) 0 B) 10 C) 12 D) 20	13. A
14.	The sum must be even, and it could be 2 + 4 + 8 =14. A) 10 B) 13 C) 14 D) 19	14. C
15.	I have 5 sweaters and 6 pairs of pants. For each sweater, there are 6 pairs of pants with which that sweater can be paired. There are 5 sweaters, so there are 5 × 6 = 30 different possible outfits. A) 11 B) 22 C) 25 D) 30	15. D
16.	Zoe is exactly twelve and one-half years old = (12×12 + 6) months = 150 months. Yolanda's age is one-third of 150 months = 50 months. A) 48 B) 50 C) 100 D) 150	16. B
17.	The smallest whole number that is divisible by 1, 2, 3, 4, 5, and 6 is 60, so the balloon man has 60 balloons. A) 21 B) 60 C) 120 D) 720	17. B
18.	My teacher uses 5 pieces of chalk every 6 days. He will use 5×5 = 25 pieces in 5×6 = 30 days. A) 25 B) 30 C) 35 D) 36	18. A
19.	Since 37 is prime and 37 + 6 = 43 is also prime, my favorite prime could be 37. A) 19 B) 29 C) 35 D) 37	19. D
20.	1 to 100 has 50 odd #s; remove the first 5. A) 45 B) 50 C) 55 D) 60	20. A
21.	I burn 4 candles each day. I need enough candles to last a week. I need 7×4 = 28 candles. Since 5 < 28÷5 < 6, I need 6 packs. A) 4 B) 5 C) 6 D) 7	21. C

Go on to the next page))))▶ **5**

2010-2011 5TH GRADE CONTEST SOLUTIONS

		Answers
22.	Each of the elephants weighs at least 3500 kg. The total weight of 5 is at least 5×3500 kg = 17500 kg. One elephant can weigh (24000 – 17500) kg. A) 4000 B) 5000 C) 5500 D) 6500	22. D
23.	He drives 120 km in 1 hour. In half an hour, he drives 60 km. In 90 minutes, he drives 180 km. A) 120 B) 150 C) 180 D) 210	23. C
24.	Even numbers between 50 and 90 include some with ones digit 0, so the product's ones digit is 0. A) 0 B) 2 C) 4 D) 8	24. A
25.	The products are 1×1, 1×2, 1×3, 1×4, 1×5, 2×1, 2×2, 2×3, 2×4, 2×5, 3×1, 3×2, 3×3, 3×4, 3×5, 4×1, 4×2, 4×3, 4×4, 4×5, 5×1, 5×2, 5×3, 5×4, 5×5. The different products are 1, 2, 3, 4, 5, 6, 8, 9, 10, 12, 15, 16, 20, and 25. A) 14 B) 17 C) 20 D) 25	25. A
26.	I have 7 coins with a value of 77¢. I have 1 quarter and 2 pennies for a total of 27¢. The other coins are 1 quarter, 1 nickel, and 2 dimes. A) 1 B) 2 C) 3 D) 4	26. B
27.	The greatest possible distance between my house and my school is 4 km + 7 km = 11 km. A) 12 km B) 7 km C) 6 km D) 3 km	27. A
28.	For each of the choices A, B, and C, the 6 whole numbers whose product is 60 are shown. A) 1,1,2,2,3,5 B) 1,1,1,3,4,5 C) 1,1,1,2,5,6 D) 17	28. D
29.	3 apples and 2 oranges cost $3.20, and 2 apples and 3 oranges cost $2.80. Adding, 5 apples and 5 oranges cost $6; 1 of each is $6÷5. A) $1.00 B) $1.10 C) $1.15 D) $1.20	29. D
30.	In 10 minutes, the amount of water in a full tub will drain, but half of it will be replaced. In 10 more minutes, the water needed to fill the tub will be added, but a full tub's worth will be drained. A) 12 B) 15 C) 20 D) 30	30. C

The end of the contest 🖐 **5**

Visit our Web site at http://www.mathleague.com

6th Grade Solutions
••••••••••••••••••••••••••
2006-2007 through 2010-2011

SIXTH GRADE MATHEMATICS CONTEST

Math League Press, P.O. Box 17, Tenafly, New Jersey 07670-0017

Information & Solutions

Tuesday, February 20 or 27, 2007

Contest Information

- **Solutions** Turn the page for detailed contest solutions (written in the question boxes) and letter answers (written in the *Answers* column to the right of each question).

- **Scores** Please remember that *this is a contest, not a test*—and there is no "passing" or "failing" score. Few students score as high as 30 points (75% correct). Students with half that, 15 points, *deserve commendation!*

- **Answers & Rating Scale** Turn to page 148 for the letter answers to each question and the rating scale for this contest.

Copyright © 2007 by Mathematics Leagues Inc.

2006-2007 6TH GRADE CONTEST SOLUTIONS

		Answers
1. $(24 + 36 + 48 + 60) \div 12 = 2+3+4+5 = 14$. A) 7 B) 9 C) 10 D) 14		1. D
2. At a quarter each, 40 gumballs will cost $40 \times 25¢ = 1000¢ = \$10.00 = \$10$. A) \$100 B) \$65 C) \$10 D) \$1		2. C
3. $45 \div 3 = 15 = 3 \times 5$. A) 3 B) 5 C) 9 D) 15		3. B
4. $1.25125 \times \dfrac{5}{4} = 1.25125 \times 1\dfrac{1}{4} = 1.25125 \times 1.25$. A) 1.20 B) 1.25 C) 1.40 D) 1.75		4. B
5. Each square's area is 9, so a side of a square is 3. The figure's perimeter is $3+9+3+9 = 24$. A) 18 B) 24 C) 27 D) 36		5. B
6. Of the following quotients, choice D is greatest. A) $64 \div 4 = 16$ B) $112 \div 7 = 16$ C) $144 \div 9 = 16$ D) $194 \div 12 > 16$		6. D
7. 450 cm = twice my height = 2×225 cm, and 225 cm $\div$ 5 = 45 cm. A) 180 cm B) 135 cm C) 90 cm D) 45 cm		7. D
8. $20 \times 40 \times 60 = 20 \times (20 \times 2) \times (20 \times 3) = 20^3 \times 6$. A) 20^1 B) 20^3 C) 20^6 D) 20^{2400}		8. B
9. 3 slices in 20 minutes $\Leftrightarrow$ $3 \times 3 = 9$ slices in 3×20 minutes = 1 hour. A) 9 B) 12 C) 15 D) 45		9. A
10. $(100 - 1) \times 99 = (100 \times 99) - (1 \times 99)$. A) 199×99 B) 198×99 C) 100×100 D) 100×99		10. D
11. Since $45 = 3 \times 3 \times 5$, choice C is correct. (Note: 1 is NOT a prime.) A) $25 = 5 \times 5$ B) $35 = 5 \times 7$ C) $45 = 3 \times 3 \times 5$ D) $55 = 5 \times 11$		11. C
12. 200% of 6 = $2 \times 6 = 3 \times 4$ = 300% of 4. A) 2 B) 3 C) 4 D) 12		12. C
13. Of 132 clowns, 66 are redheads. Of the other 66, half are blonde, so 33 are blonde. The ratio of blonde clowns to redheaded clowns is 33:66. A) 33:66 B) 44:66 C) 66:66 D) 55:77		13. A
14. "Midway" = average = $(2+12)/2 = 14/2 = 7$. A) 5 B) 6 C) 7 D) 8		14. C
15. $30\,000 + 4000 + 500 + 600 + 70 = 35\,170$. A) 345 670 B) 35 170 C) 34 170 D) 34 567		15. B

Go on to the next page ▶ **6**

2006-2007 6TH GRADE CONTEST SOLUTIONS

#	Problem	Answer
16.	The rounded value for each choice is shown below. A) 4454 B) 4450 C) 4500 D) 4000	16. C
17.	3:45 to 6:35 = 2 hrs. 50 mins. Add half that, 1 hr. 25 mins., to 3:45. A) 4:50 P.M. B) 4:55 P.M. C) 5:05 P.M. D) 5:10 P.M.	17. D
18.	For 24 683 579: hundreds' digit = 5; ten-thousands' digit = 8. A) 56 B) 40 C) 30 D) 10	18. B
19.	The average price is $(12 \times \$4 + 4 \times \$12) \div (12+4) = \$96 \div 16 = \$6$. A) \$4 B) \$6 C) \$8 D) \$12	19. B
20.	$2 \times 4 \times 6 = 8 \times 6$ is a factor in D. A) $2 \times 3 \times 4 \times 5$ B) $3 \times 4 \times 5 \times 6$ C) $4 \times 5 \times 6 \times 7$ D) $5 \times 6 \times 7 \times 8$	20. D
21.	Add until you get close: $1+2+4+8+16+32+64+128+256 = 511$. On the 9th day of May, I put down 256 grains of sand. On that day, the total number of grains put down so far that month was 511. A) May 9 B) May 10 C) May 25 D) June 19	21. A
22.	$888 \div 77 = 11$ r 41, and $41 \div 6 = 6$ r 5. The new remainder is 5. A) 0 B) 1 C) 3 D) 5	22. D
23.	101 dimes + 101 cents = \$11.11. Adding 1 cent ruins the balance. A) \$10.01 B) \$11.00 C) \$11.11 D) \$11.12	23. D
24.	$100\,001^2 - 100\,000^2 = 10\,000\,200\,001 - 10\,000\,000\,000 = 200\,001$. A) 200 001 B) 100 001 C) $200\,001 \times 10^6$ D) $100\,001 \times 10^6$	24. A
25.	Since $1+$(a number) $> 1\times$(a number), the smallest such number isn't 1. Next, $2+2 = 2\times 2$, but $2+3 < 2\times 3$, so the smallest such whole number is 2. A) 0 B) 1 C) 2 D) 3	25. C
26.	Since its diameter is 2 m, the pool's circumference is $\pi \times 2$ m. A) π m B) 2π m C) 4π m D) 5π m	26. B
27.	$\sqrt{40 \times 90} = \sqrt{3600} = 60$. A) 36 B) 50 C) 60 D) 120	27. C
28.	÷ by 2 repeatedly. If you get an odd #, there's a prime factor > 2. A) $2222 = 2\times 1111$ B) $2468 = 2^2 \times 617$ C) $4848 = 2^4 \times 303$ D) $8192 = 2^{13}$	28. D
29.	The l.c.m. of 8 and 10 is 40. That's the g.c.f. of 80 and 120. A) 160 B) 120 C) 100 D) 80	29. B

Go on to the next page ⮞ 6

#	Solution	Answer
30.	The area of each square is 16. Since $2000 \div 16 = 125$, we can fit at most 125 non-overlapping squares inside. A) 121 B) 125 C) 250 D) 500	30. B
31.	A 3-hour concert lasts 180 minutes; 18 minutes is 10% of 180 minutes. A) 6% B) 10% C) 21% D) 30%	31. B
32.	The perimeter of an equilateral $\triangle$ with integer sides is divisible by 3. A) 915 B) 615 C) 315 D) 115	32. D
33.	The 8 factors of 2006 are 1, 2, 17, 34, 59, 118, 1003, and 2006. A) 4 B) 6 C) 8 D) 9	33. C
34.	10 like both, so $20-10 = 10$ like only math, and $30-10 = 20$ like only art. That's $10+10+20 = 40$ students, so $60-40 = 20$ like neither. A) 0 B) 10 C) 20 D) 30	34. C
35.	$(1+1)+(1+3)+(1+5)+\ldots+(1+99) = 50+(1+3+5+\ldots+99) = 2550$. A) 2550 B) 2599 C) 2600 D) 5000	35. A
36.	Bob got 2, 5, 8, 11, 14, 17, ..., and Ann got 3, 5, 7, 9, 11, 13, ... Bob's 3rd, 4th, 5th, ..., 100th numbers are, respectively, 1, 2, 3, ..., 98 more than Ann's. A) 97 B) 98 C) 99 D) 100	36. B
37.	The number of leaves must be a multiple of 20. A) 399 leaves B) 400 leaves C) 401 leaves D) 410 leaves	37. B
38.	Make 1 cut through 1 thickness to get 2 pieces, 2 cuts $\Rightarrow$ 3 pieces, 3 cuts $\Rightarrow$ 4 pieces, **4** cuts $\Rightarrow$ 5 pieces, ..., 49 cuts $\Rightarrow$ 50 pieces. *At most,* I made $49 + (\mathbf{4} \times 50) = 249$ cuts. A) 249 B) 250 C) 299 D) 300	38. A
39.	Each cube needs a top and a bottom. Choice A has 2 bottoms. A) B) C) D)	39. A
40.	If 4 flips = 3 flops, and 3 flops = 6 flaps, then 4 flips = 6 flaps. A) 2 B) 3 C) 4 D) 6	40. C

The end of the contest ✏️ **6**

Visit our Web site at http://www.mathleague.com

SIXTH GRADE MATHEMATICS CONTEST

Math League Press, P.O. Box 17, Tenafly, New Jersey 07670-0017

Information & Solutions

Tuesday, February 19 or 26, 2008

Contest Information

- **Solutions** Turn the page for detailed contest solutions (written in the question boxes) and letter answers (written in the *Answers* column to the right of each question).

- **Scores** Please remember that *this is a contest, not a test*—and there is no "passing" or "failing" score. Few students score as high as 30 points (75% correct). Students with half that, 15 points, *deserve commendation!*

- **Answers & Rating Scale** Turn to page 149 for the letter answers to each question and the rating scale for this contest.

Copyright © 2008 by Mathematics Leagues Inc.

2007-2008 6TH GRADE CONTEST SOLUTIONS

		Answers
1.	Sam's age is 18. Sue's age is twice that. Her age is $2 \times 18 = 36$. A) 9 B) 27 C) 32 D) 36	1. D
2.	When 4 dozen people are paired off into couples, there are 2 dozen couples. The number of couples is $2 \times 12 = 24$. A) 12 B) 24 C) 48 D) 96	2. B
3.	9:15 A.M. to 4:11 P.M. is 7 hrs. -4 mins. Add half that, $3\frac{1}{2}$ hrs. -2 mins., to 9:15 A.M. to get 12:45 P.M. -2 mins. $= 12:43$ P.M. A) 12:43 B) 1:13 C) 1:33 D) 3:28	3. A
4.	The prime factors of $375 = 3 \times 5^3$ are 3 and 5, and $5-1 = 4$. A) 4 B) 24 C) 124 D) 374	4. A
5.	43 days ago = 6 weeks + 1 day ago = the day before Tuesday. A) Sunday B) Monday C) Wednesday D) Thursday	5. B
6.	The average side-length of a triangle with perimeter 72 is $72 \div 3$. A) 3 B) 18 C) 24 D) 36	6. C
7.	$9898 \div 3 = 3299.\overline{3}$, and $0.\overline{3} = 1/3$, so choice A has remainder 1. A) $9898 \div 3$ B) $9898 \div 4$ C) $9898 \div 6$ D) $9898 \div 7$	7. A
8.	Divide *both* numerator & denominator of $(20+30+40+50) \div 4$ by 2. A) 1 B) 2 C) 4 D) 8	8. B
9.	The number 12 345 678 910 has 6 odd digits and 5 even digits. A) 1:2 B) 2:3 C) 1:1 D) 6:5	9. D
10.	7 isn't a factor of $23 \times 24 \times 25 \times 26 = (23) \times (2^3 \times 3) \times (5^2) \times (2 \times 13)$. A) 4 B) 5 C) 6 D) 7	10. D
11.	Salty spent 45¢. Peppy began with 45¢ more than Salty. To be left with the same amount of money as Sandy, Peppy must have spent 45¢ more than Salty spent, so Peppy must have spent 90¢. A) 45¢ B) 55¢ C) 90¢ D) $1	11. C
12.	$3^2 \times 6^2 \times 9^2 = 9 \times 6^2 \times 9^2 = 6^2 \times 9^3$. A) $3^6 \times 6^3$ B) $3^6 \times 9^3$ C) $6^2 \times 9^3$ D) $6^3 \times 9^2$	12. C
13.	Cut the square into 36 1×1 squares. Each will have perimeter 4. A) 4 B) 9 C) 18 D) 36	13. D
14.	First, $3:2 = 18:12$. Of the choices listed, C is the closest to 18:12. A) 13:12 B) 15:12 C) 19:12 D) 23:12	14. C
15.	$(1+2+2+3+3+3+4+4+4+4) \div 10 = 30 \div 10 = 3$. A) 2 B) 3 C) 4 D) 5	15. B

2007-2008 6TH GRADE CONTEST SOLUTIONS

		Answers
16.	Ed began with 1 sip of milk. By his 20th sandwich bite, he had taken 10 more sips of milk. He took no more sips of milk after his 21st bite. Altogether, he took 11 sips. A) 7 B) 8 C) 11 D) 12	16. C
17.	$(111 \times 4) + (222 \times 4) = (111+222) \times 4$. A) 2 B) 3 C) 4 D) 6	17. C
18.	$(123\,123\,000\,000 + 123\,123) \div 123\,123 = 1\,000\,000 + 1 = 1\,000\,001$. A) 1 000 001 B) 1 001 001 C) 1 010 101 D) 1 111 111	18. A
19.	The product of all the factors of 12 is $(1 \times 12) \times (2 \times 6) \times (3 \times 4)$. A) 12 B) 12×12 C) $6 \times 8 \times 12$ D) $12 \times 12 \times 12$	19. D
20.	My age doubles in $4+2 = 6$ years if I turn 12 in 2 years. I'm now 10. A) 8 B) 10 C) 12 D) 14	20. B
21.	If the rectangle is a square, its area will be $5^2 = 25$. A) 20 B) 25 C) 100 D) 400	21. B
22.	8 sides + 6 sides + 5 sides = 19 sides. A) 18 B) 19 C) 20 D) 21	22. B
23.	The gcf of $(3) \times (2 \times 2) \times 5$ and $(13) \times (2 \times 7) \times (3 \times 5)$ is $2 \times 3 \times 5$. A) $3 \times 4 \times 5$ B) $2 \times 3 \times 5$ C) 3×5 D) 5	23. B
24.	The products are shown below. Only $(3/2) \times 3 = 9/2 = (3/2) + 3$. A) $\frac{2}{3} \times 3 = 2$ B) $\frac{3}{4} \times 3 = \frac{9}{4}$ C) $\frac{4}{3} \times 3 = 4$ D) $\frac{3}{2} \times 3 = \frac{9}{2}$	24. D
25.	$99\,990^2 \div 9999^2 = (9999 \times 10)^2 \div 9999^2 = (9999^2 \times 10^2) \div 9999^2 = 10^2$. A) 1 B) 10 C) 100 D) 1000	25. C
26.	The pool's center is 2 m from the pool's wall, so its radius is 2 m and its circumference is $2\pi \times 2$ m. A) π B) 2π C) 4π D) 8π	26. C
27.	It's sunny 2/3 of 36 days, or 24 days. Since I get hosed 1/4 of these 24 days, I expect to get hosed $(1/4) \times 24$ times, or 6 times, in 36 days. A) 6 B) 9 C) 12 D) 24	27. A
28.	$(25 \times 2) \times (20 \times 2) \times (15 \times 2) \times (10 \times 2) \times (5 \times 2) = (25 \times 20 \times 15 \times 10 \times 5) \times 32$. A) 2 B) 10 C) 20 D) 32	28. D
29.	100% of $10^2 = 1 \times 10^2 = 10^2 = 10 \times 10$. A) 10 B) 10^2 C) 10^3 D) 10^4	29. A

123   Go on to the next page ▶   **6**

		Answers
30.	Any number divisible by 6 and 8 is divisible by their lcm, 24, and all its factors, but not necessarily by any larger number. A) 1 B) 2 C) 3 D) 4	30. C
31.	A diameter of the largest such circle is the length of one side of the square. The square's side is $8 \div 4 = 2$, so a radius is 1. A) 8 B) 4 C) 2 D) 1	31. D
32.	A *pair* of $5 and $6 puzzles costs $11. After I buy 8 pairs for $88, I can buy 2 more $6 puzzles. A) 17 B) 18 C) 19 D) 20	32. B
33.	$\sqrt{36} \times \sqrt{36} = 6 \times 6 = 36$. A) 2×18 B) 18×18 C) 3×2 D) 36×36	33. A
34.	If my row has 26 seats, then there are 25 seats besides mine. If 5 seats are on my left and 20 are on my right, the ratio is 5:20. A) 1:1 B) 1:2 C) 1:3 D) 1:4	34. D
35.	Such products always have several factors divisible by 10, and the ones' digit of any number divisible by 10 is always a 0. A) 1 B) 5 C) 9 D) 10	35. A
36.	Subtract 1 from each term. The resulting sum is $2550 - 50 = 2500$. A) 2500 B) 2475 C) 2450 D) 1275	36. A
37.	Since 20% of 150 cm is 30 cm, and 150 cm + 30 cm is 180 cm, the Invisible Man's height without his hat on would be 180 cm − 30 cm = 150 cm. A) 144 B) 150 C) 160 D) 216	37. B
38.	5:00 to 5:59 is 60 times. At other times, 5s appear at 05, 15, 25, 35, 45, 50-59. Total # is $(60 \times 2) + (15 \times 22) = 450$. A) 362 B) 450 C) 472 D) 492	38. B
39.	A square of area 64 has side 8. Split the square into an 8×5 rectangle and an 8×3 rectangle. The smaller rectangle's perimeter is $2 \times (3+8) = 22$. A) 20 B) 22 C) 24 D) 26	39. B
40.	There are 16 △s: 6 small △s, 3 △s that use 2 small △s, 6 △s that use 3 small △s, and the original △. A) 16 B) 15 C) 13 D) 11	40. A

The end of the contest 6

Visit our Web site at http://www.mathleague.com

SIXTH GRADE MATHEMATICS CONTEST

Math League Press, P.O. Box 17, Tenafly, New Jersey 07670-0017

Information & Solutions

Tuesday, February 17 or 24, 2009

6

Contest Information

- **Solutions** Turn the page for detailed contest solutions (written in the question boxes) and letter answers (written in the *Answers* column to the right of each question).

- **Scores** Please remember that *this is a contest, not a test*—and there is no "passing" or "failing" score. Few students score as high as 30 points (75% correct). Students with half that, 15 points, *deserve commendation!*

- **Answers & Rating Scale** Turn to page 150 for the letter answers to each question and the rating scale for this contest.

Copyright © 2009 by Mathematics Leagues Inc.

2008-2009 6TH GRADE CONTEST SOLUTIONS

#	Problem	Answer
1.	$25 + 35 + 45 = (25 + 35) + 45 = 60 + 45$. A) 25 B) 35 C) 45 D) 55	C
2.	The number of weeks in 139 days is most nearly $140 \div 7 = 20$. A) 5 B) 14 C) 19 D) 20	D
3.	Together, July & August have 62 days. That's twice the 31 days in March. A) March B) April C) June D) November	A
4.	$3+6+9+12 = 30 = 3 \times 10 = 3 \times (1+2+3+4)$. A) 5 B) $1+2+3+4$ C) 12 D) $1\times 2\times 3\times 4$	B
5.	$8002 - 2008 = (8002+1) - (2008+1) = 8003 - 2009$. A) 9003 B) 9002 C) 8003 D) 8002	C
6.	If both square S and triangle T have perimeter 60 cm, then (side of T) − (side of S) = (60/3) cm − (60/4) cm = 5 cm. A) 3 cm B) 5 cm C) 8 cm D) 15 cm	B
7.	Side-length = 60/(# sides). No polygon has only 2 sides. A) $30 = 60/2$ B) $20 = 60/3$ C) $15 = 60/4$ D) $12 = 60/5$	A
8.	Average = $(\$2+\$4+\$6+\$8+\$10) \div 5 = \$6 = 600$ pennies. A) 3000 B) 600 C) 550 D) 500	B
9.	$1\times 2\times 3\times 4\times 5\times (2\times 3) = 2\times 4\times 2\times (1\times 3\times 5\times 3) = 16 \times 45$. A) 5 B) 15 C) 45 D) 75	C
10.	$50\% \times 30\%$ = half of 30% = 15% = 15% of 100%. A) 10% B) 25% C) 100% D) 150%	C
11.	To evaluate $60 \div 4 + 1 \times 3$, we first do the × and ÷ in the order in which they appear. Do the addition last. We get $15 + 3 = 18$. A) 4 B) 18 C) 36 D) 48	B
12.	$(20 \times 30) \div 40 = 600 \div 40 = 15$; remainder = 0. A) 30 B) 15 C) 10 D) 0	D
13.	$180 \div 6 = 30 = 6 \times 5$. A) 180 B) 36 C) 30 D) 5	D
14.	Ann's awake 16 hours each day. In 10 days, that's (10×16) hours. A) 10×16 B) 8×10 C) 8×24 D) 16×24	A
15.	25 dimes = 250¢ = 125×2¢ = 125×2 pennies. A) 1 penny B) 1 nickel C) 2 pennies D) 2 nickels	C
16.	1 m = 100 cm and 1 km = 1000 m; the correct ratio is 100:1000. A) 100:1000 B) 1000:100 C) 1:100 D) 100:1	A

Go on to the next page ⟶ **6**

2008-2009 6TH GRADE CONTEST SOLUTIONS

		Answers
17.	Al ran twice as far as Bob. Split 18 km into two parts, one twice the other, to see that Bob ran 6 km and Al ran twice as far, 12 km. A) 3 km B) 6 km C) 9 km D) 12 km	17. D
18.	The ones' digits repeat in groups of four: 2, 4, 8, 6, 2, 4, 8, 6, ... The ones' digits of $2^{48}, 2^{49}, 2^{50}$ are 6, 2, 4. A) 0 B) 2 C) 4 D) 8	18. C
19.	Since 225° = 180°+45°, the bird turns 45° past south. That's southwest. A) southwest B) southeast C) northwest D) northeast	19. A
20.	1 dozen = 6 pairs; 2 dozen = 12 pairs. Mult. # of dozens by 6 to get # of pairs. A) multiply, 6 B) divide, 6 C) multiply, 2 D) divide, 2	20. A
21.	Since 1+1 < 3, the side-lengths of a triangle cannot be 1, 1, 3. A) 1, 1, 3 B) 2, 2, 3 C) 3, 3, 3 D) 4, 4, 3	21. A
22.	As an example, 42 is divisible by 6 and by 14 but not by 12. A) 21 B) 12 C) 7 D) 3	22. B
23.	Rounding, 398° C is closer to 400° C than to 390° C. A) 380° C B) 390° C C) 399° C D) 400° C	23. D
24.	Increasing $1200 by 10% increases it by $120. The new price is $1320. Decreasing $1320 by 10% decreases it by $132. The final price is $1320 − $132 = $1188. A) $1212 B) $1200 C) $1188 D) $1100	24. C
25.	$10^5 + 10^6 = 1\,100\,000 = 11 \times 10^5$. A) 11 B) 12 C) 10^2 D) 10^6	25. A
26.	If 2 hoots = 1 holler, then (10 × 1) hollers = (10 × 2) hoots. A) 5 B) 13 C) 18 D) 20	26. D
27.	All but 33 can be represented as required, as shown below. A) 43 = 1+42 B) 33 C) 23 = 2+21 D) 13 = 6+7	27. B
28.	The square's perimeter = 4 × (a side) = 16. The circumference = (diameter) × $\pi \approx 6.28$. The region's perimeter is the sum ≈ 22.3. A) 3.4 B) 12.9 C) 22.3 D) 28.6	28. C
29.	$2009 = 7^2 \times 41$, and the average of 7 and 41 is $(7+41) \div 2 = 24$. A) 2009 B) 147 C) 48 D) 24	29. D

Go on to the next page ⟶ **6**

		Answers
30.	Multiply 7 by any of these: 2, 2^2, 5, 5^2, 2×5, 2^2×5, 2×5^2, or 2^2×5^2. A) 99 B) 8 C) 7 D) 5	30. B
31.	Number the coordinates 1-9, as shown. 1 2 3 4 5 6 7 8 9 The distance from B to H is $8-2 = 6$. •—•—•—•—•—•—•—•—• The distance from D to G is $7-4 = 3$. $A\ B\ C\ D\ E\ F\ G\ H\ I$ A) A to F B) E to G C) C to E D) D to G	31. D
32.	15 weeks = 105 days. If I ride for 53 of these days, then I ride 53×50 km = 2650 km. A) 2250 km B) 2600 km C) 2650 km D) 3000 km	32. C
33.	$\sqrt{8 \times 8} \times \sqrt{8 \times 8} = 8 \times 8 = 64$. A) 64 B) 32 C) 16 D) 8	33. A
34.	For every 4 girls and 1 boy, the difference is 3. The difference is *always* divisible by 3. Of the choices, only 2013 is divisible by 3. A) 2013 B) 2011 C) 2009 D) 2008	34. A
35.	2, 12, 20-29, 32, 42, 52, 62, 72, 82, and 92 use a 2; that's 19 numbers. A) 20 B) 19 C) 11 D) 10	35. B
36.	Since $806 \div 26 = 31$, the 806th letter I write will be the last letter of the 31st time I write the full alphabet; it will be a Z. A) A B) E C) V D) Z	36. D
37.	Use equivalent ratios: $3:2 = 18:12$, and $18-12 = 6$. I am now 18. In 4 years, I will be $18+4 = 22$. A) 18 B) 20 C) 22 D) 24	37. C
38.	Each of these sums is 25: $1+24$, $2+23$, $3+22$, $4+21$, $5+20$, $6+19$, $7+18$, $8+17$, $9+16$, $10+15$, $11+14$, and $12+13$. There are 12 such pairs. A) 1 B) 6 C) 12 D) 24	38. C
39.	(1 to 200) = (1 to 100) + [(100+1)+(100+2)+ ...+(100+100)] = (1 to 100) + [(100×100)+ (1 to 100)] = 5050 + [10 000 + 5050] = 20 100. A) 5150 B) 10 100 C) 11 050 D) 20 100	39. D
40.	The only possible distributions (Al,Bo,Carl) are these ten: (1,1,4), (1,2,3), (1,3,2), (1,4,1), (2,1,3), (2,2,2), (2,3,1), (3,1,2), (3,2,1), and (4,1,1). A) 10 B) 9 C) 8 D) 7	40. A

The end of the contest ✏ **6**

Visit our Web site at http://www.mathleague.com

Information & Solutions

2009-2010 Annual 6th Grade Contest

Tuesday, February 16 or 23, 2010

6

Contest Information

- **Solutions** Turn the page for detailed contest solutions (written in the question boxes) and letter answers (written in the *Answer Column* to the right of each question).

- **Scores** Please remember that *this is a contest, and not a test*—there is no "passing" or "failing" score. Few students score as high as 30 points (75% correct); students with half that, 15 points, *deserve commendation!*

- **Answers and Rating Scales** Turn to page 151 for the letter answers to each question and the rating scale for this contest.

Copyright © 2010 by Mathematics Leagues Inc.

2009-2010 6TH GRADE CONTEST SOLUTIONS

#	Solution	Answer
1.	The 3 spiders have $3 \times 8 = 24$ legs. The 3 tortoises have $3 \times 4 = 12$ legs. That's 36 legs all together. A) 14 B) 17 C) 36 D) 42	C
2.	Only 2010's digit sum is a multiple of 3. A) 2009 B) 2010 C) 2011 D) 2012	B
3.	One factor is 0, so the product is 0. A) 6400 B) 64 C) 12 D) 0	D
4.	The perimeter of a square is $4 \times$ length of a side $= 4 \times 5 = 20$. A) 10 B) 20 C) 25 D) 50	B
5.	A sum does not change when the addends are regrouped. A) $(13 + 15) + 17$ B) $(13 + 15) + (13 + 17)$ C) $(13 \times 15) + (13 \times 17)$ D) $13 \times (15 + 17)$	A
6.	The cost of 5 rides on the Ferris wheel is $5 \times 50¢ = \$2.50$. The cost of 10 rides on the roller coaster is $10 \times \$1.25 = \$12.50$. The total cost is \$15. A) \$13 B) \$14 C) \$15 D) \$16	C
7.	$\frac{1}{8} + \frac{2}{8} + \frac{3}{8} = \frac{1+2+3}{8} = \frac{6}{8} = \frac{3}{4}$. A) $\frac{3}{4}$ B) $\frac{3}{8}$ C) $\frac{3}{16}$ D) $\frac{5}{24}$	A
8.	From 8 AM yesterday until 8 AM today is 24 hours. From 8 AM till noon is 4 hours, and from noon till 3 PM is 3 hours. In all, it's $24 + 4 + 3 = 31$ hours. A) 7 B) 19 C) 31 D) 35	C
9.	$2008 + 2009 + 2010 + 2011 + 2012 = 5 \times 2010$. A) 10050 B) 10051 C) 10052 D) 10053	A
10.	Since $42 = 2 \times 3 \times 7$, 42 has 3 prime factors. A) 1 B) 2 C) 3 D) 4	C
11.	The sum of the degree-measures in *any* triangle is 180. Half of 180 is 90. A) 45 B) 90 C) 180 D) 360	B
12.	As shown below, all choices except 200 are perfect squares. A) $100 = 10^2$ B) $144 = 12^2$ C) $196 = 14^2$ D) 200	D
13.	The only common factor of any two consecutive whole numbers is 1. A) 20 B) 12 C) 2 D) 1	D
14.	The number of factors of 6 equals the exponent of 6, so we get 6^5. A) 6×5 B) 5^6 C) 6^5 D) 4^6	C

Go on to the next page 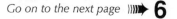 **6**

2009-2010 6TH GRADE CONTEST SOLUTIONS

		Answers
15.	Amy's age is three times her little sister Bo's age, so Bo is 18 ÷ 3 = 6. Since Charles' age is three times the sum of Amy's and Bo's ages, Charles' age is 3 × (18 + 6) = 3 × 24 = 72. A) 54 B) 60 C) 66 D) 72	15. D
16.	0.1 × 0.2 = 0.02 = 1/50, and 1/50 × 250 = 5. A) 1000 B) 530 C) 500 D) 250	16. D
17.	There are 5 in each block of 10 up to 2091. There are 4 more up to 2099. A) 44 B) 45 C) 88 D) 89	17. A
18.	The average of five equally-spaced numbers is the middle number. A) 85 B) 85.5 C) 86 D) 86.5	18. C
19.	If 6 students are wearing jeans, then 18 − 6 = 12 are not. The ratio of students wearing jeans to students *not* wearing jeans is 6:12 = 1:2. A) 1:2 B) 1:3 C) 2:3 D) 2:1	19. A
20.	The sum of 2 numbers is 12, and their product is 35. The numbers are 5 and 7. The larger of the two numbers is 7. A) 8 B) 7 C) 6 D) 5	20. B
21.	Since 123 is divisible by 3, and (8+9+10+11) = 38 is even, 3 × 2 is a factor. A) 9 B) 8 C) 7 D) 6	21. D
22.	Since twice the perimeter of a square, tripled, is 72, the perimeter is (72 ÷ 3) ÷ 2 = 12. One side's length is 12 ÷ 4 = 3, so the square's area is 9. A) 3 B) 9 C) 12 D) 16	22. B
23.	Choice C is correct as shown below. A) $1^5 = 1$ B) $2^4 = 16$ C) $3^3 = 27$ D) $4^2 = 16$	23. C
24.	May has 16 odd-numbered and 15 even-numbered days. Dave ran 16 times for 15 min. and 15 times for 44 min. That's a total of (15×16)+(15×44) = 15×(16 + 44) min. = 15 hours. A) 15 B) 30 C) 60 D) 900	24. A
25.	5×√5×5×√5 = 5×5×√5×√5 = 5×5×5. A) 5 × 5 × 25 B) 5 × 5 × 5 C) 5 × 5 × 2 D) 5 × 5	25. B
26.	The product of two whole numbers is 30. If the numbers are 5 and 6, their sum is 5 + 6 = 11. A) 10 B) 11 C) 13 D) 31	26. B
27.	222 × 66 = (2 × 111) × (2 × 3 × 11) = (3 × 111) × (2 × 2 × 11) = 333 × 44. A) 1 B) 2 C) 3 D) 4	27. A

Go on to the next page ⟫ **6**

2009-2010 6TH GRADE CONTEST SOLUTIONS

#	Problem	Answer
28.	Each number is added 3 times and subtracted once, so B is correct. A) (8+10+12) B) 2 × (8+10+12) C) 3 × (8+10+12) D) 4 × (8+10+12)	B
29.	The number 123 is a whole number between 100 and 999 that has three different non-zero digits; the sum of its digits is 1 + 2 + 3 = 6. A) 7 B) 6 C) 4 D) 3	B
30.	Ed is 31 − 20 = 11, and Di is 35 − 20 = 15. The sum of their ages is 11 + 15 = 26. A) 26 B) 46 C) 86 D) 106	A
31.	Since 1000 ÷ 12 has R4, it's 4 months after Mar. A) March B) May C) June D) July	D
32.	Multiply the ones digits: 3 × 6 × 9 = 162. A) 1 B) 2 C) 3 D) 4	B
33.	The value of one of each coin is (1+5+10)¢ = 16¢. Since $2.40÷16¢ = 15, there are 15 of each coin. The value of 15 nickels is 15 × 5¢ = 75¢. A) 15¢ B) 50¢ C) 75¢ D) 95¢	C
34.	Each difference is 5. There are 2010÷5 fives = 402 fives = 5^{402}. A) 5^{402} B) 5^{401} C) 5 × 402 D) 5 × 401	A
35.	Two equilateral triangles share sides with a square as shown. The figure has 6 sides of length 4, so the perimeter is 6 × 4 = 24. A) 48 B) 40 C) 32 D) 24	D
36.	There are 420 students in my school. The ratio of boys to girls in my school *cannot* be 11:14 since 11 + 14 = 25 is not a factor of 420. A) 3:7 = 126:294 B) 5:9 = 150:270 C) 11:14 D) 17:18 = 204:216	C
37.	3×300 = 900, and 900÷3000 = 0.3 = 30%. A) 10 B) 25 C) 30 D) 50	C
38.	See choices. One of each brick weighs 10 kg. Subtract 10 repeatedly from each choice until the difference is 0 or divisible by 3 or 7. A) 21 kg B) 27 kg = 2×3+3×7 C) 30 kg = 3×3+3×7 D) 39 kg = 6×3 + 3×7	A
39.	If (1×2×3×…×30)+1 is divided by 2 or 3 or 5 or … or 29, the remainder is always 1. A) less than 10 B) between 10 & 20 C) between 20 & 30 D) greater than 30	D
40.	Each block, 1–99, 100–199, 200–299, 400–500, has 10 such numbers. From 300 to 399, there are 100−10 = 90 numbers. In all, there are 40 + 90 = 130 numbers. A) 130 B) 140 C) 150 D) 160	A

The end of the contest 6

Visit our Web site at http://www.mathleague.com

Information & Solutions

2010-2011 Annual 6th Grade Contest

Tuesday, February 15 or 22, 2011

6

Contest Information

- **Solutions** Turn the page for detailed contest solutions (written in the question boxes) and letter answers (written in the *Answer Column* to the right of each question).

- **Scores** Please remember that *this is a contest, and not a test*—there is no "passing" or "failing" score. Few students score as high as 28 points (80% correct); students with half that, 14 points, *deserve commendation!*

- **Answers and Rating Scales** Turn to page 152 for the letter answers to each question and the rating scale for this contest.

Copyright © 2011 by Mathematics Leagues Inc.

2010-2011 6TH GRADE CONTEST SOLUTIONS

		Answers
1.	Subtract the 4 different letters used in "apple" from the 26 letters in the alphabet to get 22. A) 22 B) 21 C) 20 D) 19	1. A
2.	The remainder when the sum 2011 + 201 + 20 + 2 is divided by 10 is 1 + 1 + 0 + 2 = 4. A) 1 B) 2 C) 4 D) 8	2. C
3.	222×4 + 222×2 + 222×1 = 222×(4+2+1). A) 3 B) 6 C) 7 D) 8	3. C
4.	Three of four parents are mothers. Therefore, three–fourths of the 60 parents are mothers, and (3/4) × 60 = 45. A) 15 B) 20 C) 40 D) 45	4. D
5.	66 + 34 + 66 + 34 + 66 + 34 + 66 + 34 + 66 + 34 + 66 + 34 + 66 + 34 + 66 + 34 + 66 + 34 + 66 + 34 = (66 + 34) × 10 = 100 × 10 = 1000. A) 1000 B) 900 C) 660 D) 100	5. A
6.	The digit sum is 21, and 4 divides 56, so 123 456 is divisible by 3, 4, and 6. A) 3 B) 4 C) 6 D) 7	6. D
7.	The product of two odd numbers is odd, so there is only one even product, 2 × 11. A) 0 B) 1 C) 2 D) 3	7. B
8.	A rectangle has 4 right angles. Any 2 adjacent sides are perpendicular. A) rectangle B) rhombus C) trapezoid D) parallelogram	8. A
9.	(1 × 2) + (3 × 4) + (5 × 6) + (7 × 8) = 2 + 12 + 30 + 56 = 100. A) 36 B) 100 C) 1256 D) 4680	9. B
10.	In 2011, February has 28 days. So 1 week = 7 days is 7/28 = 0.25 = 25% of the 28 days in February. A) 4% B) 7% C) 25% D) 35%	10. C
11.	Rachel has $200 in her piggy bank. Last year the amount in her piggy bank was 20% of $200 = $40. A) $20 B) $40 C) $80 D) $160	11. B
12.	If, 8 years ago, Sid was as old as Sue was 2 years ago, then Sid is 6 years older than Sue. Their present ages are 18 and 12. A) 12 B) 18 C) 22 D) 28	12. B
13.	$6 \times 6^{2011} = 6^1 \times 6^{2011} = 6^{2012}$. A) 6^{2012} B) 6^{2016} C) 36^{2011} D) 36^{2016}	13. A

Go on to the next page))))▶ 6

2010-2011 6TH GRADE CONTEST SOLUTIONS

	Answers
14. The number of multiples of 3 from 33 to 66 is the same as the number from 3 to 36. There are 12 bus lines in Betaburg. A) 14 B) 13 C) 12 D) 11	14. C
15. Two numbers with an average of 10 have a sum of 20. Since 7×13 = 91, the smaller is 7. A) 7 B) 9 C) 11 D) 13	15. A
16. 1 + 2 + 4 + 8 + 16 + 32 = 63. A) 7 B) 24 C) 30 D) 63	16. D
17. 24 hours = 1440 minutes, and 1440 ÷ 5 = 288, so it drips 288 times. A) 36 B) 72 C) 144 D) 288	17. D
18. The 4 small rectangles all have equal areas. The combined areas of 3 of these rectangles is 12, so each has an area of 4. The area of the the original rectangle is 16. A) 4 B) 16 C) 18 D) 48	18. B
19. The average of 21, 23, 25, … , 47, and 49 is (21 + 49) ÷ 2 = 35. A) 33 B) 34 C) 35 D) 36	19. C
20. The greatest prime number less than 40 is 37, and the least prime number greater than 20 is 23. Their difference is 37 − 23 = 14. A) 14 B) 16 C) 18 D) 19	20. A
21. Forty minutes after 4:40 is 5:20, and 5:20 is forty minutes before 6:00. A) 5:20 B) 6:00 C) 6:20 D) 6:40	21. B
22. $9^{201} \times 10^{201} = 90^{201} = 6^{201} \times 15^{201}$. A) 13^{201} B) 15^{201} C) 17^{201} D) 19^{201}	22. B
23. 0.1 × 0.2 = 0.02 and 0.02 × 1500 = 30. A) 30 B) 300 C) 600 D) 1500	23. D
24. The g.c.d. of 6 × 7 × 8 × 9 and 7 × 8 × 9 × 10 is (7 × 8 × 9) × (the g.c.d. of 6 and 10). A) 7×8×9 B) 2×7×8×9 C) 6×7×8×9 D) 7×8×9×10	24. B
25. If I round 12 345 to the nearest 100, I get 12 300. Then, 12 300 ÷ 1230 = 10. Finally, 10 × 123 = 1230. A) 12340 B) 1234 C) 1230 D) 123	25. C

Go on to the next page 6

2010-2011 6TH GRADE CONTEST SOLUTIONS

		Answers
26.	9 tiles cover a 6 × 6 section, leaving a 1-unit wide uncovered area. Pat needs 7 more tiles to cover this 1-unit wide area. A) 13 B) 14 C) 15 D) 16	26. D
27.	From 1 to 999 are 249 multiples of 4. Remove the 25 multiples from 500 to 599 and the 18 others that end in 52 or 56. A) 206 B) 207 C) 208 D) 209	27. A
28.	The ratio 3:5 is equal to $(10 \times 3):(10 \times 5) = 30:50$. A) $\frac{1}{3}:\frac{1}{5}$ B) 13:15 C) 30:50 D) $(1+3):(1+5)$	28. C
29.	My number could be 16 since the sum of 16's factors is $1+2+4+8+16 = 31$, which is less than twice 16. A) 12 B) 16 C) 20 D) 24	29. B
30.	Once around, we go up 3, so we must go down 3. We must go right 1+2+3, and we must go left 3+2+1. The total distance is 3+3+6+6 = 18. A) 18 B) 20 C) 21 D) 22	30. A
31.	Choose any distance from Brad's to Caryn's house. Let's use 35 km, a number divisible by 5 and 7. It takes Brad 35÷5 = 7 hrs. and 35÷7 = 5 hrs. for the trips. His avg. speed is $(d \text{ km})/(t \text{ hrs.}) = 70/(5+7) = 35/6$. A) $5\frac{5}{7}$ B) $5\frac{5}{6}$ C) 6 D) $6\frac{1}{6}$	31. B
32.	Factoring: $(1 \times 2 \times 3 \times \ldots \times 48) \times (49 \times 50 - 49) = (1 \times 2 \times 3 \times \ldots \times 47 \times 48) \times 49^2$. A) 49^2 B) 50^2 C) 49 D) 50	32. A
33.	Player 1 plays 9 games, then leaves. Player 2 plays 8 other games, then leaves, player 3 plays 7 other games, then leaves, and so on. The total number of games is 9+8+7+6+5+4+3+2+1 = 45. A) 100 B) 90 C) 50 D) 45	33. D
34.	Subtract 6 from each choice to find the number of cats with spots. We get 2, 3, 12, and 18. The number of cats without spots for each is 24, 23, 14, and 8. Since 3×8 = 24 = 18+6, it's choice D. A) 8 B) 9 C) 18 D) 24	34. D
35.	The decimal is 0.0185185.... An "8" appears in the 3rd, 6th, 9th, … , 2010th decimal place. So a "5" is in the 2011th place. A) 0 B) 1 C) 5 D) 8	35. C

The end of the contest 6

Visit our Web site at http://www.mathleague.com

Answer Keys & Difficulty Ratings

2006-2007 through 2010-2011

ANSWERS, 2006-07 4th Grade Contest

1. C	7. D	13. C	19. B	25. B
2. D	8. B	14. C	20. A	26. C
3. D	9. B	15. A	21. C	27. A
4. A	10. A	16. A	22. A	28. B
5. A	11. B	17. B	23. C	29. C
6. D	12. D	18. D	24. D	30. B

RATE YOURSELF!!!
for the 2006-07 4th GRADE CONTEST

Score	Rating
28-30	Another Einstein
25-27	Mathematical Wizard
23-24	School Champion
19-22	Grade Level Champion
17-18	Best In The Class
15-16	Excellent Student
12-14	Good Student
9-11	Average Student
0-8	Better Luck Next Time

ANSWERS, 2007-08 4th Grade Contest

1. D	7. B	13. C	19. B	25. D
2. B	8. A	14. A	20. D	26. C
3. B	9. D	15. A	21. D	27. C
4. A	10. C	16. B	22. A	28. B
5. C	11. A	17. D	23. B	29. A
6. C	12. D	18. C	24. A	30. D

RATE YOURSELF!!!
for the 2007-08 4th GRADE CONTEST

Score	Rating
28-30	Another Einstein
24-27	Mathematical Wizard
22-23	School Champion
19-21	Grade Level Champion
17-18	Best In The Class
14-16	Excellent Student
12-13	Good Student
10-11	Average Student
0-9	Better Luck Next Time

ANSWERS, 2008-09 4th Grade Contest

1. C	7. D	13. C	19. A	25. B
2. B	8. A	14. C	20. A	26. A
3. A	9. A	15. C	21. C	27. D
4. C	10. D	16. D	22. B	28. A
5. C	11. D	17. B	23. D	29. D
6. B	12. B	18. B	24. C	30. B

RATE YOURSELF!!!
for the 2008-09 4th GRADE CONTEST

Score	Rating
29-30	Another Einstein
26-28	Mathematical Wizard
23-25	School Champion
21-22	Grade Level Champion
19-20	Best In The Class
16-18	Excellent Student
13-15	Good Student
10-12	Average Student
0-9	Better Luck Next Time

ANSWERS, 2009-10 4th Grade Contest

1. D	7. D	13. D	19. A	25. D
2. C	8. A	14. C	20. B	26. B
3. B	9. B	15. C	21. B	27. C
4. D	10. A	16. D	22. D	28. D
5. D	11. B	17. D	23. A	29. A
6. C	12. A	18. A	24. B	30. C

RATE YOURSELF!!!
for the 2009-10 4th GRADE CONTEST

Score	Rating
29-30	Another Einstein
27-28	Mathematical Wizard
24-26	School Champion
22-23	Grade Level Champion
19-21	Best In The Class
17-18	Excellent Student
14-16	Good Student
11-13	Average Student
0-10	Better Luck Next Time

ANSWERS, 2010-11 4th Grade Contest

1. A	7. A	13. A	19. C	25. A
2. C	8. D	14. B	20. B	26. B
3. D	9. C	15. D	21. C	27. A
4. B	10. C	16. D	22. C	28. C
5. B	11. B	17. C	23. B	29. B
6. A	12. B	18. C	24. D	30. D

RATE YOURSELF!!!
for the 2010-11 4th GRADE CONTEST

Score	Rating
29-30	Another Einstein
26-28	Mathematical Wizard
24-25	School Champion
22-23	Grade Level Champion
19-21	Best In The Class
17-18	Excellent Student
14-16	Good Student
11-13	Average Student
0-10	Better Luck Next Time

ANSWERS, 2006-07 5th Grade Contest

1. C	7. D	13. A	19. B	25. D
2. D	8. B	14. B	20. D	26. C
3. C	9. B	15. C	21. D	27. B
4. C	10. D	16. C	22. C	28. D
5. A	11. C	17. A	23. A	29. A
6. A	12. D	18. C	24. B	30. B

RATE YOURSELF!!!
for the 2006-07 5th GRADE CONTEST

Score	Rating
28-30	Another Einstein
26-27	Mathematical Wizard
23-25	School Champion
21-22	Grade Level Champion
19-20	Best In The Class
17-18	Excellent Student
14-16	Good Student
12-13	Average Student
0-11	Better Luck Next Time

ANSWERS, 2007-08 5th Grade Contest

1. C	7. D	13. A	19. B	25. B
2. A	8. B	14. A	20. A	26. B
3. A	9. B	15. C	21. B	27. A
4. A	10. C	16. B	22. C	28. D
5. B	11. D	17. C	23. D	29. C
6. B	12. D	18. D	24. C	30. D

RATE YOURSELF!!!
for the 2007-08 5th GRADE CONTEST

Score	Rating
29-30	Another Einstein
27-28	Mathematical Wizard
24-26	School Champion
22-23	Grade Level Champion
19-21	Best In The Class
16-18	Excellent Student
14-15	Good Student
11-13	Average Student
0-10	Better Luck Next Time

ANSWERS, 2008-09 5th Grade Contest

1. C	7. B	13. A	19. C	25. B
2. D	8. B	14. D	20. B	26. C
3. C	9. A	15. B	21. A	27. D
4. A	10. C	16. A	22. D	28. B
5. D	11. B	17. B	23. D	29. D
6. A	12. C	18. C	24. D	30. A

RATE YOURSELF!!!
for the 2008-09 5th GRADE CONTEST

Score	Rating
29-30	Another Einstein
27-28	Mathematical Wizard
24-26	School Champion
21-23	Grade Level Champion
18-20	Best In The Class
15-17	Excellent Student
13-14	Good Student
10-12	Average Student
0-9	Better Luck Next Time

ANSWERS, 2009-10 5th Grade Contest

1. C	7. A	13. B	19. A	25. A
2. C	8. A	14. D	20. D	26. C
3. B	9. B	15. B	21. C	27. D
4. D	10. A	16. D	22. A	28. B
5. A	11. D	17. A	23. A	29. D
6. C	12. C	18. B	24. B	30. D

RATE YOURSELF!!!
for the 2009-10 5th GRADE CONTEST

Score	Rating
28-30	Another Einstein
26-27	Mathematical Wizard
24-25	School Champion
21-23	Grade Level Champion
19-20	Best In The Class
16-18	Excellent Student
13-15	Good Student
11-12	Average Student
0-10	Better Luck Next Time

ANSWERS, 2010-11 5th Grade Contest

1. C	7. B	13. A	19. D	25. A
2. A	8. A	14. C	20. A	26. B
3. D	9. B	15. D	21. C	27. A
4. A	10. C	16. B	22. D	28. D
5. C	11. D	17. B	23. C	29. D
6. B	12. A	18. A	24. A	30. C

RATE YOURSELF!!!
for the 2010-11 5th GRADE CONTEST

Score	Rating
28-30	Another Einstein
26-27	Mathematical Wizard
24-25	School Champion
22-23	Grade Level Champion
19-21	Best In The Class
17-18	Excellent Student
14-16	Good Student
12-13	Average Student
0-11	Better Luck Next Time

ANSWERS, 2006-07 6th Grade Contest

1. D	9. A	17. D	25. C	33. C
2. C	10. D	18. B	26. B	34. C
3. B	11. C	19. B	27. C	35. A
4. B	12. C	20. D	28. D	36. B
5. B	13. A	21. A	29. B	37. B
6. D	14. C	22. D	30. B	38. A
7. D	15. B	23. D	31. B	39. A
8. B	16. C	24. A	32. D	40. C

RATE YOURSELF!!!
for the 2006-07 6th GRADE CONTEST

Score	Rating
38-40	Another Einstein
35-37	Mathematical Wizard
32-34	School Champion
28-31	Grade Level Champion
25-27	Best In The Class
21-24	Excellent Student
18-20	Good Student
15-17	Average Student
0-14	Better Luck Next Time